ATLAS OF AFRICA

This innovative *Atlas of Africa*, by the Agence Française de Développement, offers comprehensive insights into contemporary Africa through the use of full-colour maps, charts, graphics and text which demonstrate and explain Africa's growing importance in the world and its demographic, economic, social and environmental transformation, while also outlining the challenges that the continent faces.

The three sections, offering new perspectives on the continent, comprise:

- Taking Full Measure of Africa—examining the major economic, demographic, social and political transformations that Africa has undergone in a short space of time.

- A Multifaceted Continent with Shared Challenges—looking at the major intraregional economic, demographic, environmental and social dynamics that are currently shaping the continent.

- Africa Inventing Itself and Taking Up the Key Challenges of Tomorrow—an overview of the challenges that Africa is currently facing and will need to face in the future, including the environment and climate change, social cohesion and demographic issues, economic development and governance.

Full-colour maps, charts and graphics cover such wide-ranging topics as economic development, urbanization, education, the rule of law, gender, the blue economy, regional organizations, energy and culture, to form a volume which offers a thorough overview in graphic form of Africa in the world today, of interest to all those studying, working in or with Africa, and those with a general interest in the continent.

The **AGENCE FRANÇAISE DE DÉVELOPPEMENT (AFD)**, a public financial institution, funds, supports and accelerates the transition towards a fairer and more sustainable world. Its teams, focusing on climate, biodiversity, peace, education, urban development, health and governance, carry out projects in France's overseas departments and territories and in a further 115 countries.

ATLAS OF AFRICA

NEW PERSPECTIVES ON THE CONTINENT

PREFACE BY VERA SONGWE
FOREWORD BY RÉMY RIOUX

LONDON AND NEW YORK

First published 2021
by Routledge
2 Park Square, Milton Park, Abingdon, Oxon OX14 4RN

and by Routledge
605 Third Avenue, New York, NY 10158

Routledge is an imprint of the Taylor & Francis Group, an informa business

© 2022 Agence Française de Développement (AFD)

Originally published in France as:
Atlas de l'Afrique AFD. Pour un autre regard sur le continent
By AGENCE FRANÇAISE DE DEVELOPPEMENT
© Armand Colin 2020, Malakoff
© English Translation, AFD 2020
ARMAND COLIN is a trademark of DUNOD Editeur - 11, rue Paul Bert - 92240 MALAKOFF.

The right of the AFD to be identified as author of this work has been asserted by them in accordance with sections 77 and 78 of the Copyright, Designs and Patents Act 1988.

All rights reserved. No part of this book may be reprinted or reproduced or utilised in any form or by any electronic, mechanical, or other means, now known or hereafter invented, including photocopying and recording, or in any information storage or retrieval system, without permission in writing from the publishers.

Trademark notice: Product or corporate names may be trademarks or registered trademarks, and are used only for identification and explanation without intent to infringe.

British Library Cataloguing-in-Publication Data
A catalogue record for this book is available from the British Library

Library of Congress Cataloging-in-Publication Data
Names: Agence frandcaise de dâeveloppement, creator. | Songwe, Vera, writer of preface. | Rioux, Râemy, writer of foreword.
Title: Atlas of Africa : new perspectives on the continent / Agence Francaise de Developpement ; preface by Vera Songwe ; foreword by Remy Rioux.
Other titles: Altas de l'Afrique AFD. English
Identifiers: LCCN 2021010457 (print) | LCCN 2021589062 (ebook) | ISBN 9781032038438 (hardback) | ISBN 9781032039756 (paperback) | ISBN 9781003190028 (ebook) | ISBN 9781003190028q (ebook) | ISBN 9781032038438q (hardback) | ISBN 9781032039756q (paperback)
Subjects: LCSH: Africa--Maps. | Economic development--Africa. | Geopolitics--Africa--Maps. | Africa--Economic conditions--21st century--Maps. | Africa--Social conditions--21st century--Maps. | LCGFT: Atlases.
Classification: LCC G2445 A44 2021 (ebook) | LCC G2445 (print) | DDC 912.6--dc23
LC record available at https://lccn.loc.gov/2021589062
LC record available at https://lccn.loc.gov/2021010457

ISBN: 978-1-032-03843-8 (hbk)
ISBN: 978-1-032-03975-6 (pbk)
ISBN: 978-1-003-19002-8 (ebk)
DOI: 10.4324/9781003190028

Typeset in Franklin Gothic
by Taylor & Francis Books

CONTENTS

VII **PREFACE**

IX **ALL OF AFRICA**
For a new vision of a continent on the move

XI **PREAMBLE**
Africa takes on Covid-19

XIII **ABBREVIATIONS**

TAKING FULL MEASURE OF AFRICA

2   The prominence of Africa's demographic equilibria around the world

4   Longer life expectancy and a gradual reduction of fertility rates

6   Africa, the home of the world's youth: an opportunity to be seized

8   Child and maternal health and under-nutrition: gaining the upper hand

10   Contagious and tropical diseases: Africa exposed but headed in the right direction

12   Africa's efforts in schooling its children

14   Water and sanitation: the key to improving health and environmental conditions

16   Africa on the path to electrical connectivity

18   The success of mobile telephones facilitates inclusive banking

20   African economic growth despite instability

22   Africa's asset: the internal market

24   A changing economy pursuing its own development path

26   Closer economic ties with the rest of the world

28   Africa, the new travel destination

30   The courting of Africa at the diplomatic and cultural level

32   Political liberalization in Africa

34   Less lethal conflicts, more complex violence

36   The gradual improvement in economic and public governance

A MULTIFACETED CONTINENT WITH SHARED CHALLENGES

- 40 African giants and small economies: how to read the African economy
- 42 The geography of wealth in Africa
- 44 Considerable natural resources: who for? what for?
- 46 Extreme poverty concentrated on the continent
- 48 Fragility and development: two interconnected challenges
- 50 Inequalities: a situation of contrasts
- 52 Urbanization, a strong trend in settlement in Africa
- 54 Simultaneous densification of cities and countryside
- 56 Rural and landlocked zones: the main factors in unequal access to basic services
- 58 Economic integration progressing in Africa
- 60 Growth in intra-African trade as a driver of diversification
- 62 Emergence of a pan-African financial system
- 64 Intra-continental migration: human ties between African areas
- 66 Human mobility taking shape on the continent
- 68 Creative industries: the face of an emerging cultural industry
- 70 Africa, a blue continent
- 72 The Saharan-Sahelian space
- 74 African forests under pressure

AFRICA REINVENTING ITSELF AND TAKING UP THE KEY CHALLENGES OF TOMORROW

- 78 A continent that is vulnerable to climate change
- 80 The impacts of climate change
- 82 African biodiversity under pressure
- 84 Skills acquisition and enhancement: the education challenge in Africa
- 86 Employment in all its forms: a key to the prosperity of the continent
- 88 Issues concerning women's status in Africa
- 90 Obesity and chronic diseases: new health challenges
- 92 High expectations for the rule of law and freedom of expression
- 94 Inventive civil society in Africa
- 96 Sport as a vector of economic and social development
- 98 Solar energy in Africa: reconciling sustainability and economic opportunities
- 100 Towards more inclusive African cities
- 102 Expanding internet access to accelerate the digital revolution
- 104 Building the capacity for an active public sector
- 106 The potential of the public development banks in Africa
- 108 Towards more bank financing for the private sector
- 110 Reinventing external financial support for Africa
- 112 Agriculture the African way
- 114 Industrialization: drawing benefit from value chains
- 116 What role could digital innovation play in Africa?

118 INDEX

PREFACE

The structural changes taking place around the world in the 21st century have come to represent a geostrategic challenge for development on the African continent and have given fresh impetus to the transformation of Africa into a world power, building on the spirit of Pan-Africanism, its history of resilience and today's African renaissance. A new awareness and the efforts undertaken since the turn of the millennium have culminated in the reassertion by the Heads of State and of Government of their commitment to shape a Pan-African vision of an "integrated, prosperous, and peaceful Africa, driven by its own citizens and representing a dynamic force in the international arena." The African Union's "The Africa We Want" Agenda 2063 was adopted in 2015. It envisions a long-term, 50-year development path and describes the way in which the continent intends to achieve its ambitions by attaining inclusive growth and sustainable development. Through these aspirations, the peoples of Africa are expressing their desire for shared prosperity and well-being, unity and integration, for a continent of free citizens with broader horizons, where women and young people, boys and girls alike, can realize their full potential and be sheltered from fear, disease and need.

Despite great development efforts on the continent, there are still many hindrances slowing down its economic rise. Clichés, stereotypes and preconceived perceptions remain, but this narrative provides only one version of the facts. The reality is that Africa is endowed with a remarkable abundance of natural and human resources and is driven by positive demographic and economic trends which, if managed in a way that is fair and viable, can turn the situation around over time and place the continent on the path to inclusive growth and sustainable development. The time has therefore come to change perceptions and work with Africa differently.

This Atlas proposes a dynamic insight into the sustainable development challenges facing Africa, addressed on a global scale. By combining the most relevant data, it delves into subjects that most reference books rarely touch upon. It fulfills a decisive role in furthering economic, social and sustainable progress in Africa, on the one hand by adopting a continental approach taking account of all its diversity and complexity, and on the other by providing a new vision of an Africa on the move, reviewing the main trends observed over the past few decades and identifying key challenges for the future. It thus provides a better understanding of the profound and fast-moving transformations at work on the continent, and also of its diversity, challenges and prospects.

The analyses and illustrations provided here on the main aspects of African development underscore the absolute necessity of changing our perceptions if we are to achieve a finer understanding of the rise and renewal of the continent. They offer new insights into Africa and the way in which it is upending demographic equilibria around the world, pursuing its economic growth and building a diversified and prosperous economy that is both vast and resilient, in a complex process that is at once political, economic, social and cultural. Encouraging reforms continue to bring substantial progress in terms of peace, governance and access to basic social services (education, health, water, energy, infrastructure and ICT, etc.). The continent's economic, human and cultural potential, coupled with improved governance and processes of democratization and economic integration, such as the African Continental Free Trade Area (AfCFTA), along with the development of tourism and new information technologies, are opening up new opportunities to address the challenges of fragility, population growth, poverty, inequality, rapid urbanization and vulnerability to climate change.

At present, Africa is also dealing with the coronavirus (Covid-19) pandemic, the most severe worldwide health crisis of our times and the greatest challenge we have faced since the Second World War. In a show of regional and international solidarity, the African countries have mobilized their forces to deal with the health and socioeconomic emergencies caused by the epidemic. Whatever setbacks there might be along the way, however, our determination to shape "The Africa We Want" is unwavering.

Drawing on our successful experience in managing other epidemics and working within the framework of a broad strategic partnership, Africa will organize a coordinated, effective and urgent response to this shock, steering the continent back on its path to growth and development and, finally, boosting its resilience in order to achieve the Sustainable Development Goals (SDGs) and the objectives of Agenda 2063. Of course, as is already the case, this will require greater coordination of our collective action at the highest possible level, the mobilization of multiple stakeholders, including the private sector and civil society, and a sense of purpose and urgency.

This *Atlas of Africa* is the fruit of cooperation between the United Nations Economic Commission for Africa (ECA) and the Agence Française de Développement (AFD), and will contribute to further raising awareness among development actors of the new win-win strategic partnership that "a prosperous Africa" needs in order to carve out its place in the world economy. Today, Africa is asserting its position in the world as a key partner, opening up new possibilities by capitalizing on its assets and potential in all their diversity.

VERA SONGWE
UNDER SECRETARY-GENERAL OF THE UNITED NATIONS AND EXECUTIVE SECRETARY OF THE ECA

ALL OF AFRICA

FOR A NEW VISION OF A CONTINENT ON THE MOVE

This Atlas was born out of an apparent paradox: although Africa is the closest continent to our own, it is without doubt the one we are least familiar with and, all too often, the most caricatured. The reason for this is a thought process based on patterns that no longer have any foundation, such as dividing the continent up into two parts, North Africa and Sub-Saharan Africa. Such approaches perpetuate a vision of Africa invented by the Western world and do not do justice to the emergence and coherence of the continent as a whole, as demonstrated once again by the coronavirus crisis. If we are to change our view, then it is an "All of Africa" approach that must come to the fore. That is the rationale behind this book.

It provides a dynamic vision with a contemporary insight into a continent on the move in these times of Sustainable Development Goals (SDGs). While revealing the major challenges still to be taken up in Africa, brought to light once again by the Covid-19 crisis (in matters of access to good-quality education, equality between men and women, and designing sustainable infrastructures and cities, to name but a few), this publication also offers a view of a continent that is taking shape in every area, from entrepreneurship to key infrastructure and from culture to sport. A continent driven by its young people who are enterprising, innovative and inspirational for the rest of the world.

Above all, it is an atlas that provides a very concrete vision of Africa, drawing on the realities of the projects supported by the Agence Française de Développement (AFD) and conducted by a community or a country, an NGO or a municipality, whether locally, nationally or regionally. It derives its pedagogical force from a series of illustrations chosen by my colleagues at the AFD, examples that give physical form to our vision, and without which it might prove too abstract. This exhaustive analysis of development, accompanied by a variety of examples as a complement to the general information, provides the most direct understanding of the nature and magnitude of the changes underway.

In order to put this narrative of an Africa that is taking shape into images and figures, the AFD has also carried out considerable work compiling data and maps. This is an unprecedented effort insofar as there is not yet any such thing as a statistical Africa—often for lack of reliable data—and also because what statistics there are still treat North Africa and Sub-Saharan Africa differently. The data contained in this publication are therefore the result of unprecedented statistical collection and aggregation work, making it all the more original and contributing to much-needed research efforts on contemporary Africa.

We have tried to make these graphs, maps and texts lively and dynamic, very much in the image of the continent itself, and they have been put together to form three parts providing insight into Africa as a whole. First of all, to understand and underscore the major transformations that Africa has succeeded in making at high speed, and which are often ignored. Without underestimating the challenges that remain, Africa has already proven its incredible adaptability ("Taking Full Measure of Africa"). Then, to demonstrate that it is the scale of the continent as a whole that is the relevant level for observation, including when seeking to decipher the variety of dynamics at work within it. The multiple spaces that compose it are thus interconnected by the dynamics that are shaping Africa today ("A Multifaceted Continent with Shared Challenges"). And finally, to take a look, without any defeatism or illusions whatsoever, at the upcoming challenges and the responses that are being prepared. That is the message of the final, forward-looking part of the publication: to show more clearly how Africa, along with the rest of the world and more particularly with Europe and France, is renewing itself ("Africa Inventing Itself and Taking Up Tomorrow's Challenges").

Finally, I would like to express my warmest thanks to my friend Vera Songwe, the formidable Executive Secretary of the United Nations Economic Commission for Africa, who is ever present and active in the crisis and who agreed to write the preface to this book. Along with the teams, she is the finest observer and analyst of African transitions. I would also like to express all my gratitude to Editions Armand Colin for believing in this project from the outset and sharing their precious publishing expertise with us. Finally, I would like to commend the great work carried out by the valiant AFD team in charge of the publication, headed by Clémence Vergne and Christophe Cottet, who put all their skills and passion into the project.

Because we are convinced that "It is in Africa that part of the world's changeover will be played out," to quote French President Emmanuel Macron in his speech in Ouagadougou on 28 November 2017. And because we want to build a world in common.

RÉMY RIOUX
CHIEF EXECUTIVE OFFICER OF AGENCE FRANÇAISE DE DÉVELOPPEMENT

PREAMBLE

AFRICA TAKES ON COVID-19

This *Atlas of Africa* is published at a historic moment in time. We are experiencing troubles the like of which have not been seen in peacetime since the 1930s: the threat of economic crisis has appeared almost simultaneously around the world in the wake of a slump in demand caused by the decline in international trade and lockdown measures introduced by governments to limit the spread of coronavirus. Although it is still difficult, at this stage, to quantify the medium- to long-term structural impacts of this crisis, the short-term consequences are considerable. The world economy is set to experience its deepest recession since the Great Depression, after activity contracted sharply in 2020.

Like other regions around the world, Africa is being hit hard by the Covid-19 crisis. However, many African states have demonstrated their ability to react very promptly to the crisis. Almost all of them introduced measures to close borders or impose lockdowns right from the start of the crisis, encouraged by the consistent political position of the African Union. In addition to this fast reaction, mention should also be made of Africa's long experience in managing epidemics, as shown recently in the cases of AIDS and then Ebola. However, the shock suffered by the continent is still likely to prove to be unprecedented, economically and socially. In 2020, Africa experienced the first recession in its recent history, due to the collapse in world trade and the fall in prices for most commodities. In this new crisis of globalization, the African economy is vulnerable, for the continent is largely dependent on the rest of the world for its sourcing and exports. The crisis could also have serious social consequences. The World Bank predicted that poverty rates would rise in Africa in 2020, bringing an end to the downward trend seen over the past 20 years. The United Nations Development Programme (UNDP) is concerned about the drop in human development indicators, as the crisis is hitting particularly hard in the areas of education (school closures, early school leaving, little online learning) and health. The crisis is also hitting hard among the African SMEs that have emerged over the past 20 years. It is thus amplifying inequalities.

The nature of this crisis requires a government stimulus policy to try to offset the fall in demand by providing support to the population and businesses. However, unlike in the advanced economies, the capacity of the African states and their central banks to implement this sort of "countercyclical" policy is limited on the whole (low government revenues, limited debt capacity, limited monetary financing). Identifying the funding needed to support the continent in its transformation is therefore more critical than ever. This crisis calls for a strong international response, guided by the need for solidarity and shared responsibilities, in which public development banks must play an important role. It requires national and regional public funding tools to be developed to mobilize local savings to finance a part of the recovery without increasing the debt levels of the African states. Indeed, the 80 African development banks have been quick to act, emphasizing once again the key role these players have in times of crisis.

The crisis can also be a driver of change for Africa, in a number of ways. First of all, it could open up a new debate on the role of the state, notably in welfare protection and in implementing structural reforms over the long term. It can act as an accelerator of decision-making to give priority to certain sectors that are all too often neglected (health, welfare systems, digital technologies, governance). Second, the crisis has revealed the vulnerability of globalized value chains. Onshoring of certain activities within the region could enable the African continent to diversify and boost demand, create integrated production sectors, and develop innovation capacities in sectors such as renewable energy and industry, for instance, but only if we show ourselves capable of guiding and financing these reinvestment decisions. Finally, the crisis demonstrates more clearly than ever the need for coordinated action among African countries for the benefit of their continent, with a decisive role to be played by the regional and Pan-African organizations, such as the United Nations Economic Commission for Africa or the African Union.

Ultimately, the Covid-19 crisis is highlighting issues that already existed on the continent. These great challenges for the future are many (shifting economic models towards greater diversification, the impacts of climate change, pressures on biodiversity, protecting human capital, governance issues, reinforcing social ties, etc.). This Atlas provides a detailed insight into these key challenges, based on analysis and experience, and shows the continent's capacity for innovation. Far from a monolithic vision of Africa, it proposes nuanced analyses that take account of all the diversity of the processes at play regionally and nationally, at an unprecedented moment in history.

ABBREVIATIONS

AIDS	acquired immunodeficiency syndrome
CO_2	carbon dioxide
Dem.	Democratic
etc.	et cetera
GDP	gross domestic product
i.e.	id est (that is to say)
km	kilometre(s)
HIV	human immunodeficiency virus
m	metre(s)
ND	Not defined/No data
NGO	non-governmental organization
p.	page
Rep.	Republic
SMEs	small and medium-sized enterprises
sq	square
TWh	terrawatt hour(s)
UN	United Nations
UNESCO	United Nations Educational, Scientific and Cultural Organization

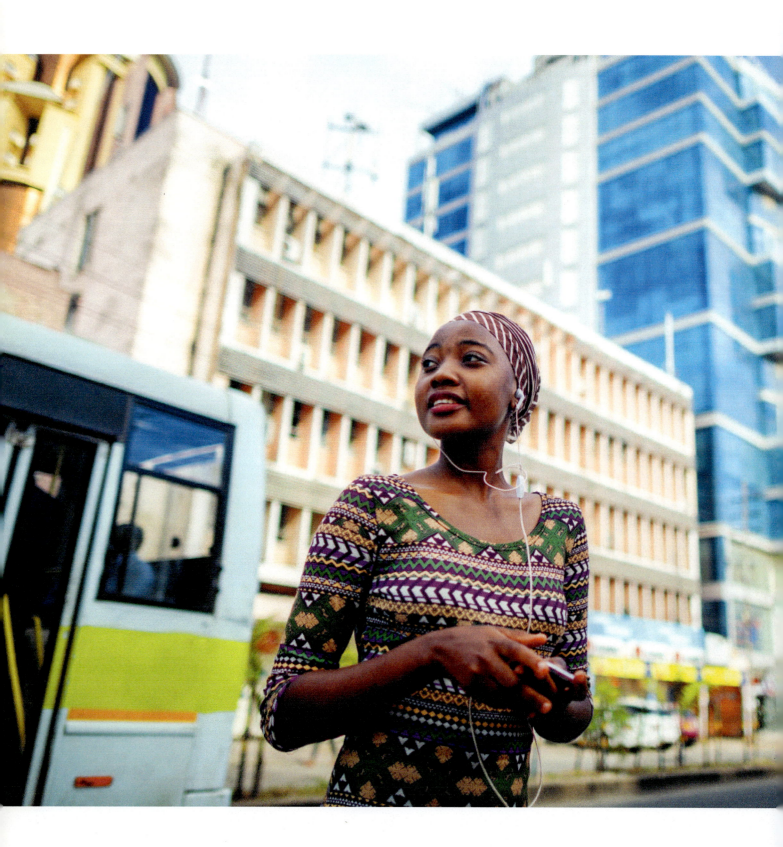

TAKING FULL MEASURE OF AFRICA

Africa's place in the world has grown in recent years, as the center of gravity of the key global issues has begun to shift towards the continent, and the African countries have moved quickly to diversify their economic, political and financial partners. Africa, this close neighbor of Europe, has already undertaken transformations on a scale never seen before in the history of the continent, and sometimes in the history of the world. These profound and fast-moving changes concern all the areas of development, whether economic, political, demographic or social. Although little noticed, many of them are bringers of good news. Of course, some of the old challenges are still there at a time when new ones are emerging, and certain countries may prove to be in a stronger position than others when it comes to addressing them, but what this first part illustrates above all else is Africa's ability to adapt.

Young woman in the city centre of Dar es Salaam, Tanzania.
Photo © wilpunt/Getty Images.

The prominence of Africa's demographic equilibria around the world

RECORD POPULATION GROWTH

Africa is in the throes of a population surge: between 1980 and 2018, its population increased by 168% (an average annual rate of 2.6%), from 476.4 million to 1.3 billion people. In the medium variant of the United Nations population projections, the African population will probably reach 2.5 billion people in 2050 and 4.3 billion by 2100. The speed and scale of this current trend are unprecedented.

In Africa, the increase from 1 to 2 billion people will take 29 years (between 2009 and 2038), whereas the same increase in Asia was spread over 40 years (between 1928 and 1968). However, Africa may take longer than Asia did to reach the 3- and then 4-million thresholds. For Africa this is 25 years (2038–2063) and then 27 years (2063–2090), respectively, compared to 19 years for Asia (1968–1987 and then 1987–2006).

A MAJOR PLACE IN WORLD DEMOGRAPHICS TODAY

As the last continent to experience this population growth phenomenon—at a time when it is slowing down everywhere else—Africa is set to completely transform the distribution of the world's population. Since 2000, it has been the only continent to see its share of the world population increase significantly.

Although Africans represented 13% of the total world population at the end of the 20th century, they are likely to make up one-fourth of the population by 2050, and 40% by 2100. By the end of the 21st century, Africa will thus be home to several billion individuals and will be beginning to challenge Asia for the top spot as the most populated region on Earth. The fate of Africa's population will therefore be increasingly the fate of the world population as a whole.

POPULATION SHIFTING ACROSS THE CONTINENT

The continent's human geography is also being changed. Six of the ten countries that will make the biggest contribution to world population growth through to 2050 will be African: Nigeria, the Democratic Republic of the Congo, Ethiopia, the United Republic of Tanzania, Egypt and Uganda. Meanwhile, the two regions of Africa that concentrate the wealthiest countries and those that are the most advanced in their demographic transition (North Africa and Southern Africa) will see their share of the African population decline gradually.

In contrast, the countries around the Gulf of Guinea and in East Africa, the two most densely populated regions on the continent, will see their numbers grow to represent half the African population by 2050. It is the Sahel that will see the strongest population growth, however, as the countries there see an increase in their share in the African population to 10% by 2050.

A group of schoolchildren playing with a globe. Norton, Zimbabwe, 2015.
Photo © CECIL BO DZWOWA/Shutterstock.

Demographic weight of Africa in the world in 2050

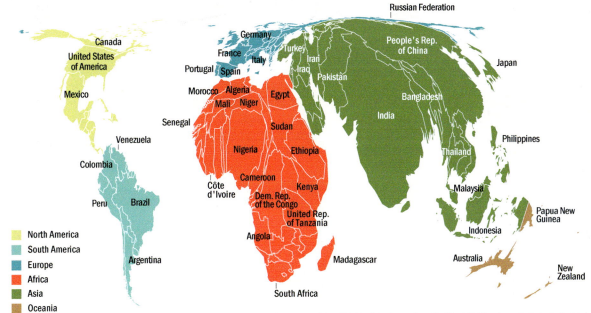

Source: United Nations (World Population Prospects) annd World Bank (World Development Indicators for South Sudan).

The size of countries is proportional to the number of inhabitants in 2050.

Demographic trends on the continent

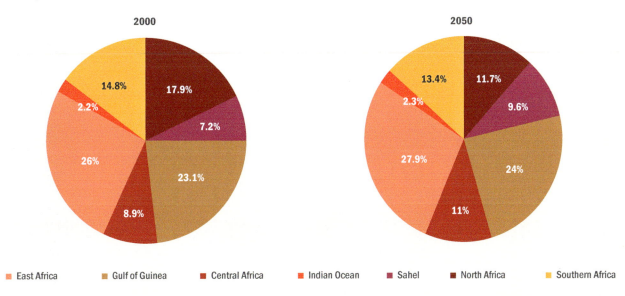

■ East Africa ■ Gulf of Guinea ■ Central Africa ■ Indian Ocean ■ Sahel ■ North Africa ■ Southern Africa

Source: United Nations (World Population Prospects).

Longer life expectancy and a gradual reduction of fertility rates

A VICTORY IN HEALTH DRIVING POPULATION CHANGE

How can we explain the exceptional demographic growth being experienced by Africa at present? First of all, it is down to a victory in health and development: people are dying later and later in Africa. Life expectancy has thus increased by 25 years since the 1950s, to 63 years, which is just nine years below the worldwide average of 72 years. This progression in life expectancy, although a little slower than in Asia, is still exceptional, and is one of the main factors driving the increase in the population on the continent. Most importantly, according to UN projections, life expectancy should continue to rise as the living conditions and health of Africans improve, and should be practically on a par with the current longevity level in Europe by the end of the 21st century, at 78 years.

FERTILITY RATES STILL HIGH BUT ON THE WAY DOWN

At the same time, the number of children per woman is still high across the continent as a whole: the fertility rate in 2017 stood at 4.5 children per woman, which is the highest in the world. However, it is on the way down; between 1980 and 2017, it fell from 6.6 to 4.5 children per woman.

This average rate conceals significant differences across the continent: the countries of Southern Africa and North Africa have completed or are on the point of completing their demographic transition, with fertility rates that are down to between three and four children per woman. The number of children per woman is decreasing in all the regions of Africa, however. In the Sahel, for instance, the region with the highest fertility rates, the number of children per woman has dropped from 7.2 to 5.7 since 1980.

A GRADUAL FALL IN POPULATION GROWTH

Africa therefore finds itself in this very particular situation in which life expectancy is increasing sharply while fertility is declining more gradually, thereby causing the particularly prolific population growth of the last few decades and probably of those to come. Asia and Latin America experienced a similar phase

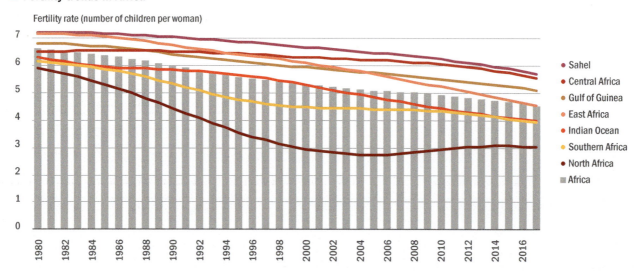

Fertility trends in Africa

Source: World Bank (World Development Indicators).

4 • ATLAS OF AFRICA

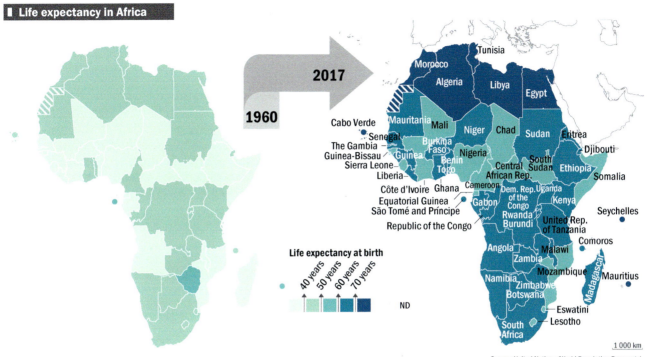

Life expectancy in Africa

Life expectancy at birth: 40 years, 50 years, 60 years, 70 years. ND

Source: United Nations (World Population Prospects).

at the end of the 20th century and as in these regions in the past, it will not last forever.

Although population growth may be higher in Africa than in the rest of the world, the fall in the population growth rate—the pace at which the population is growing—is already underway.

However, it has been falling at a slightly slower pace than that observed so far in other regions of the world.

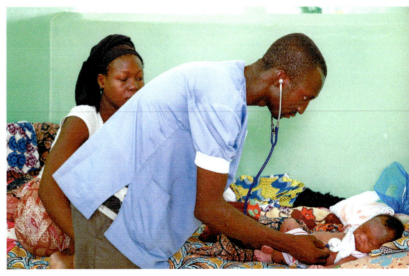

This young mother gave birth to her little boy prematurely. Since the birth, he has made a strong recovery and the trainee pediatrician is checking that he is now fit to leave the hospital. The "pediatrics package" introduced as part of the Chad Health Sector Support Project (PASST2) provides comprehensive coverage for children from birth to five years old.

Photo: Joseph Barbereau © Clotilde Bertet/AFD.

Africa, home to the world's youth: an opportunity to be seized

ALREADY THE WORLD'S YOUNGEST CONTINENT...

Africa is already the world's youngest continent, and is set to remain so in the coming decades. With almost 800 million people under the age of 25 out of a total population of 1.3 billion, the proportion of young people in the population stood at 60% in 2020, compared to 39% in Asia and 26% in Europe, and the under-25s will represent over half of the African population at least until 2050.

The age structure of the African population is therefore very close to that in pre-1960 People's Republic of China, characterized by a large share of young people in the demographic structure and an age pyramid that is weighted heavily towards its base (see graph).

... AND SOON HOME TO HALF OF THE WORLD'S YOUTH

As the rest of the world gradually ages, Africa will be home to a growing proportion of the planet's youth. By 2050, more than one-third of young people between the ages of 15 and 24 and 40% of the world's children will be African. From 2070 onwards, the number of young Africans under the age of 25 should reach 1.5 billion, representing almost half of the world's youth. Across the continent, however, the regions in the north and south will have started to age, as they have already begun their demographic transition. The countries in these regions will therefore have to cope with the challenges of financing their retirement pension systems, and they need to begin preparing today. It is in the countries in the middle of Africa that the youngest populations will be found: West and East Africa, and in Central Africa.

Age structure of the population: comparison between Africa (2019, left) and China (1950, right)

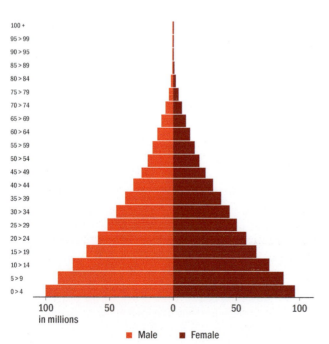
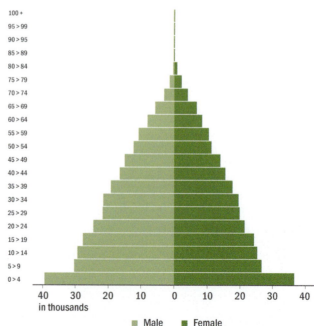

Source: United Nations (World Population Prospects).

In Isare, a rural area on the outskirts of Bujumbura (Burundi), this youth centre provides school and extra-curricular courses for the children of the municipality, thanks to the *Ideas Box* of Libraries Without Borders. Photo © Kibuuka Mukisa Oscar/AFD.

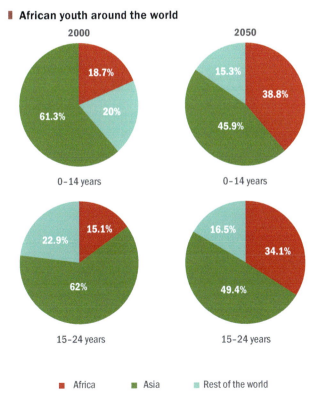

CREATING THE CONDITIONS FOR A DEMOGRAPHIC DIVIDEND

This process can constitute an opportunity for Africa, as it opens up a potential "demographic dividend". Today, there are too few Africans of working age in relation to the number of children and elderly people. Education, health, welfare and retirement pensions are costly and take up a large part of the income of the active population.

This situation is set to evolve, however, as today's children become adults who are active in the labour market. Africa's working population as a proportion of the total population is expected to be higher than that in Europe and North America by 2040, and the highest in the world from about 2060 onwards. Financing welfare expenditure will therefore be easier. This will be the moment when the continent benefits from a "demographic dividend", a phenomenon that is currently underway in Asia. The challenge for Africa will be to create the overall conditions to benefit from this potential by investing in education and health, and by creating a dynamic labour market capable of absorbing the active population (see pp. 84–85 and 86–87).

TAKING FULL MEASURE OF AFRICA • 7

Child and maternal health and under-nutrition: gaining the upper hand

CHILD MORTALITY FALLING RAPIDLY

One of the most positive aspects of the population trends on the continent is the considerable decline in child mortality. This has been divided by three in about 40 years: while almost one child in five under the age of five died in 1980 in Africa, the proportion had been brought down to one in 15—equivalent to a rate of 66 in 1,000—by 2018.

This progression has become increasingly generalized throughout the continent: while 85% of African countries had a child mortality rate in excess of 100 per 1,000 children at the beginning of the 1980s, just six of them still have such levels today. This progress has gone hand-in-hand with a considerable fall in maternal mortality, which has almost halved.

EXPANSION IN BASIC CARE FOR MOTHERS AND CHILDREN

This encouraging result has been achieved at the cost of considerable efforts by the African states and the international community in favour of child and maternal health. The proportion of women receiving pre-natal care has risen from 65% in 2000 to 83% in 2018, and the number of births assisted by qualified healthcare personnel from 48% to 63%.

Africa has also benefited from both public or private health plans on a large scale, notably in vaccination. The action of the Global Alliance for Vaccines and Immunization (GAVI), which is cofinanced by donors and countries from all over Africa, has been an important factor in the progression in vaccination rates among children in Africa. The share of children aged 12–23 months that have received the Diphtheria, Tetanus, Pertussis (DTP3) vaccine, for example, has increased from 9.8% in 1980 to 79.3% in 2018, which places Africa at the same level as Europe in the 1980s.

UNDER-NUTRITION RATES NEED TO BE REDUCED

In one other area, Africa has succeeded in keeping away the threat of hunger that marked it so severely in the 1970s and 1980s: the prevalence of under-nutrition has thus declined from 28.3% of the population in 1991 to 21.1% in 2014. This rate has shown an unfortunate upturn recently, mainly due to conflicts, and mainly those in the Sahel countries bordering Lake Chad and in the Horn of Africa: it thus reached 22.4% in 2017.

After wars, global warming is the second structural cause of under-nutrition, as drought hits harvests and livestock. However, the continent is neither the only or first region in the world for hunger: the number of people affected by hunger is 1.7 times lower there than in Asia and in 2017, one-third of the 811 million under-nourished people on earth lived in Africa.

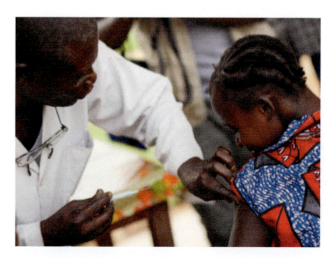

Vaccination campaign against tetanus, Democratic Republic of the Congo.
Photo © Valeriya Anufriyeva/Shutterstock.

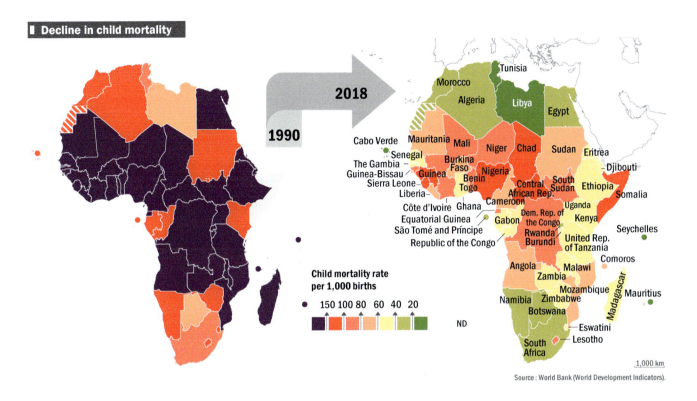

Decline in child mortality

1990 → 2018

Child mortality rate per 1,000 births: 150 100 80 60 40 20 ND

Source: World Bank (World Development Indicators).

Vaccination around the world (2018)

Percentage of children vaccinated against DTP*
- 90 %
- 80 %
- 60 %
- 40 %
- ND

* Diphtheria, Tetanus, Pertussis.

Source: World Bank (World Development Indicators).

TAKING FULL MEASURE OF AFRICA

Contagious and tropical diseases: Africa exposed but headed in the right direction

A GROWING EPIDEMIC RESPONSE CAPACITY

On account of their tropical climates and fragile health systems, the African countries are regularly hit by outbreaks of epidemics. Overall, however, the trend is towards a reduction in transmissible and infectious diseases. Africa has benefited from progress in medicine and immunology via access to the treatments and vaccines developed in the most advanced countries. The incidence of most of the major contagious diseases—tuberculosis, polio, measles—has thus fallen all over the continent. The same goes for HIV, of which more than 65% of sufferers live in Africa, but which has seen its prevalence and incidence (number of cases identified and of new cases) decline constantly since 2000.

The Covid-19 pandemic has also shown that a large majority of the countries on the continent were able to react rapidly and decisively to the appearance of the new virus, notably via early prevention measures (travel restrictions and curfews). The recurrence of epidemics—Ebola in West Africa in 2014 and 2015 was one of the most recent—also facilitates adoption of good practice (detection, isolation of people with the disease, precautions in basic care and hygiene, such as hand washing).

THE CURSE OF TROPICAL DISEASES: A CHALLENGE TO BE TAKEN UP, BUT PROGRESS ON MALARIA

On account of its geographic position and climate conditions, the continent is still home to endemic diseases for which there is no treatment. These "neglected tropical diseases" (NTDs), as they are called, because they hit poor populations that do not constitute a market for the pharmaceutical industry, include little-known pathologies (schistosomiasis, leprosy, sleeping sickness, etc.) which, along with malaria, are thought to cause more deaths each year in Africa than tuberculosis.

Malaria is caused by a parasite that is spread by mosquitoes and is now concentrated mainly on the African continent: 92% of the world's cases identified in 2017 were in Africa, causing the deaths of over 400,000 people that year.

However, unlike the NTDs, progress in antimalarial treatments and enhanced prevention efforts, especially via use of mosquito nets, have driven a reduction in mortality from this disease (see graph). It has therefore fallen by almost 30% in Africa, from 533,000 deaths in 2010 to 380,000 deaths in 2018.

PREVENTION AS A BASIC TREATMENT

If Africa is taking more time to wipe out infectious diseases that have disappeared or almost disappeared elsewhere, it is mainly due to the shortcomings of its healthcare systems which suffer all over the continent from budgets that are too small and significant gaps in their facilities and human resources.

The lack of hospitals and affordable drugs and shortages of qualified personnel, estimated to amount to at least 5.6 million health professionals, limit not only access to treatments, but also the scope of prevention efforts.

Yet, prevention is less costly than therapeutic medicine and in certain cases can constitute the key to reducing pathologies, especially those that are caught through social practices that can be modified (diet, sexuality, funeral rites, contraception, clothing habits, etc.).

This social dimension of health makes human resources a key factor in human development in Africa, at a time when the trend is rather towards a flight of qualified individuals from leaving the continent.

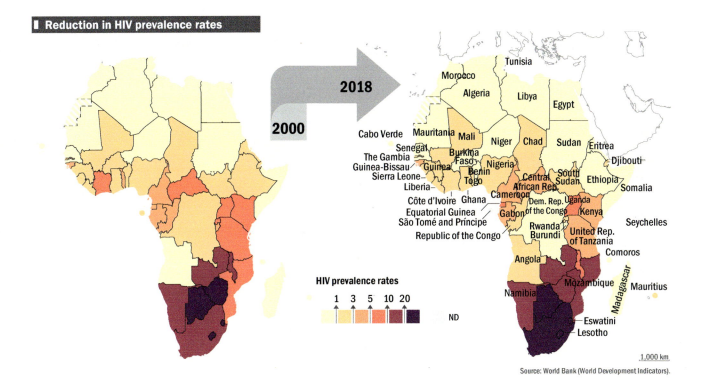

Reduction in HIV prevalence rates

2000 → 2018

HIV prevalence rates: 1, 3, 5, 10, 20, ND

Source: World Bank (World Development Indicators).

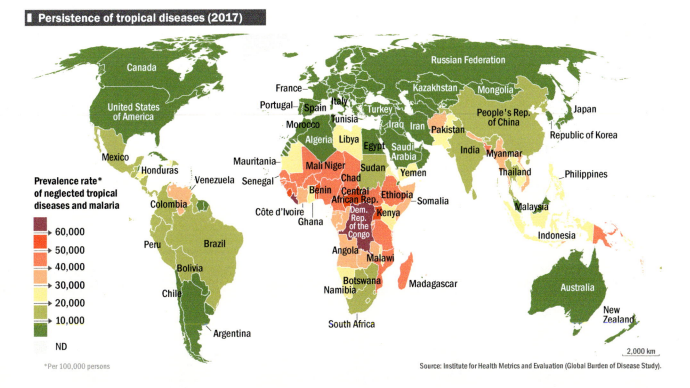

Persistence of tropical diseases (2017)

Prevalence rate* of neglected tropical diseases and malaria:
- →60,000
- →50,000
- →40,000
- →30,000
- →20,000
- →10,000
- ND

*Per 100,000 persons

Source: Institute for Health Metrics and Evaluation (Global Burden of Disease Study).

Africa's efforts in schooling its children

SUSTAINED PROGRESSION IN SCHOOL ENROLMENT RATES

In the space of barely 30 years, a discreet revolution has taken place in education in Africa. Despite a sharp and rapid increase in the number of children, the African states have succeeded in increasing the proportion of those enrolled in primary schools from 55% in 1980 to 78% in 2018. This represented some 180 million children attending primary schools, which is close to 100 million more than in 1990.

The effects of this large-scale school enrolment policy have been immediate: while in 1990, African adults had spent an average of three years at school, three decades later they had been in schooling for six years. The proportion of African adults who are able to read and write thus increased from 49% to 66% between 1995 and 2018.

A SHARED PUBLIC PRIORITY

These trends owe nothing to chance: the progress achieved in access to basic education is chiefly the fruit of an investment by the African states. The share of public expenditure dedicated to education is among the highest in the world in many cases.

On average, education represents 18.5% of government spending on the continent, against 14% in the world as a whole (see map). On average, Africa devotes almost 4% of its GDP to education, a higher level than that in South Asia (3.4%) and East Asia (3.6%), but less than in Latin America (4.5%) (see graph). Families also dedicate a large share of their budgets to educating their children, with some of them enrolling them at private schools.

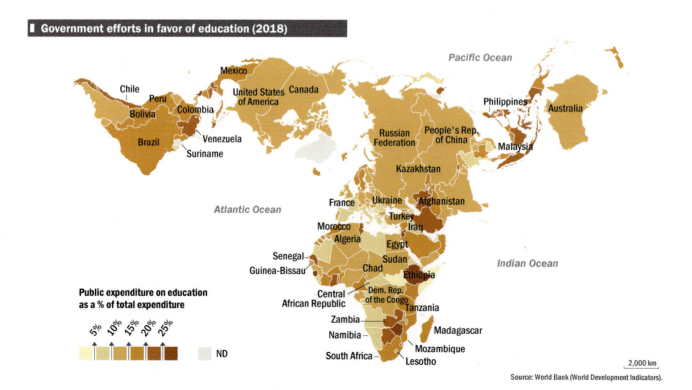

Government efforts in favor of education (2018)

Source: World Bank (World Development Indicators).

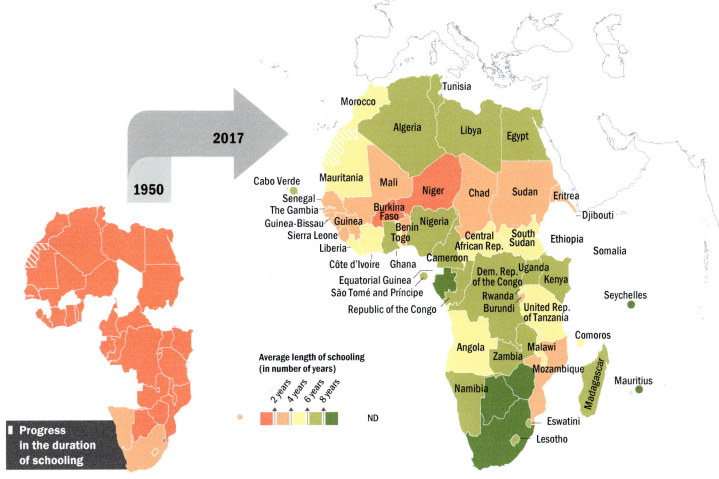

Progress in the duration of schooling

Average length of schooling (in number of years): 2 years, 4 years, 6 years, 8 years, ND

Source: Our World in Data (Barro and Lee).

INEQUALITIES IN ACCESS TO EDUCATION

The battle for schooling is not over yet, however, and in Africa more so than elsewhere. In 2018, 34 million African children of primary school age were not in school and the rapid increase in the number of children requires school capacities to be boosted further. About two-thirds of those children around the world who are not enrolled in school live in Africa.

Greater efforts need to be made in some regions. In the countries of North Africa, almost all children are enrolled in primary education, but only just over two-thirds of children living in the Sahel and the Gulf of Guinea are enrolled. Schooling for girls is also a challenge of decisive importance if the general situation is to be improved. At the level of the continent, the rate of enrolment in primary education among boys was 80% in 2018, and that figure was higher than that for girls, at 76%. This objective of enhanced inclusion must be pursued at the same time as other issues in education, and in particular extending the duration of school education and improving its quality—two great challenges for the future (see pp. 84–85).

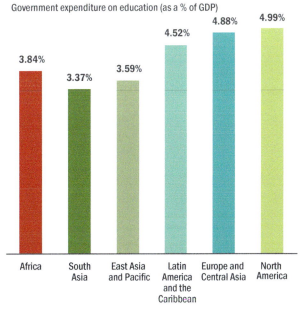

Budget allocated to education (2017)

Government expenditure on education (as a % of GDP)

- Africa: 3.84%
- South Asia: 3.37%
- East Asia and Pacific: 3.59%
- Latin America and the Caribbean: 4.52%
- Europe and Central Asia: 4.88%
- North America: 4.99%

Source: World Bank (World Development Indicators).

Water and sanitation: the key to improving health and environmental conditions

VARYING RATES OF IMPROVEMENT IN ACCESS TO ELEMENTARY POTABLE WATER AND SANITATION

There is another area that is crucial for the well-being and health of the population on the continent, and in which Africa has made progress: access to drinking water. A growing number of Africans have access to at least a basic water supply service, meaning equipped water points, such as protected wells, drillholes or standpipes. The share of the population with access to this service has thus increased from 53% in 2000 to 66% in 2017.

Although the situation differs between the regions of Africa, the improvement in this access to water—by 24% between 2000 and 2017—has been the highest in the world overall in recent years.

However, the progress achieved in sanitation has been slower: the share of Africans with access to elementary sanitation facilities (meaning adequate toilets) increased only from 28% to 33% between 2000 and 2017.

EFFORTS TO BE CONTINUED

This relative improvement in access to water and sanitation must be stepped up. First, because one person in two without access to a basic water service lives in Africa. That represents 413 million people, which is 38 million more than in 2000 due to population growth and despite the efforts that have been made.

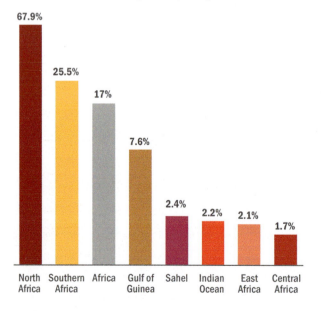

■ Access to sanitation systems (2017)

Source: UNICEF and World Health Organization (2019).

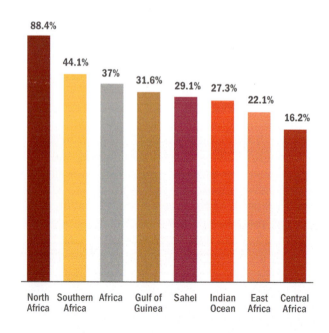

■ Access to drinking water at home (in sufficient quantity and quality, 2017)

Source: UNICEF and World Health Organization (2019).

This is all the more true in sanitation, on account of the sheer scale of the investments required, the large number of actors involved, and institutional and financial frameworks that are still fragile in many cases.

Next, because access to higher-quality services is still low in all cases: only 37% of the population of Africa has access to water of sufficient quality and in sufficient quantity at home, while only 17% have sanitation systems that are managed in perfect safety, meaning that there are adequate toilets and wastewater and sludge treatment facilities.

A CHALLENGE FOR HEALTH AND THE ENVIRONMENT

Access to drinking water and sanitation represents a public health challenge for Africa. Given the lack of suitable infrastructures and hygiene practices, the prevalence of water-borne diseases such as cholera and dysentery remains high. Poor sanitation is thought to be the cause of almost two-thirds of deaths from diarrhea on the continent In addition, re-use of untreated wastewater for agricultural purposes is a widespread practice that constitutes a risk of contamination of foodstuffs in the city, while stagnant water is conducive to swarms of mosquitoes, including those that transmit malaria.

Another advantage of developing sanitation systems, re-using treated water and optimizing water use (reducing leakage on networks, etc.) is to achieve better management of a resource that is becoming increasingly scarce. Africa is partly exposed to aridity and water stress (when potable water requirements exceed available resources, in particular in the north and the south of the continent), a situation that is being compounded by the effects of climate change. Reinforcing water governance in accordance with integrated management on a water-basin basis to guarantee coordinated distribution of resources between users (farming, domestic, industrial) is therefore also to be encouraged (see pp. 70–71).

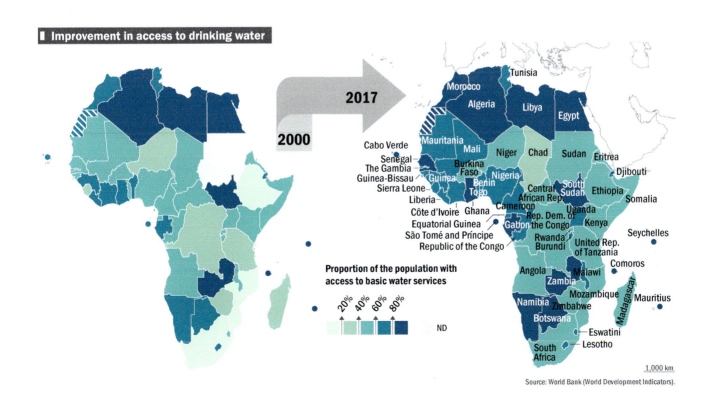

Improvement in access to drinking water

Source: World Bank (World Development Indicators).

TAKING FULL MEASURE OF AFRICA • 15

Africa on the path to electrical connectivity

ACCESS TO ELECTRICITY: AFRICA ENGAGED IN A FAST-TRACK CATCH-UP PROCESS

Africa has made major progress in terms of electrification. The share of the population with access to electricity jumped from 29% to 53% between 1990 and 2017, bringing almost 470 million extra Africans the benefits of this energy source. The result is that Africa now enjoys comparable electrification levels to those in South Asia at the start of the 2000s. Access to electricity is thus one of the areas in which Africa is catching up with the rest of the world. Even if this access does remain well below the world average (89% of the world population in 2017), the rate of progress is much more sustained here (see graph).

PROGRESSING OVERALL, BUT SUBSTANTIAL DIFFERENCES

The other stand-out fact in the progression in access to electricity in Africa is the genuine improvement throughout the continent: almost all the African countries recorded an increase in access between 1990 and 2017, some of them at a particularly fast pace. For instance, Kenya saw its access to electricity grow from 3% to 64% of the population in the space of less than 30 years. Yet despite this, there is still a long way to go to achieve universal connection in all the territories of the continent. Over 575 million Africans are still deprived of electricity and access remains uneven from one country to another.

Generally speaking, access to electricity is higher in the countries along the African coastline, and lower in the inland countries in the heart of the continent. In certain countries, the rate of access to electricity remains very low, such as in Burundi, where only 9% of the population enjoys access to it. Significant differences also exist within countries, in particular between the cities, which are better connected, and the countryside where access is much more limited on average (see pp. 56–57).

INNOVATION, THE KEY TO ELECTRIFICATION

The progress made so far and that yet to be made is no doubt driven by increasing investments in the electricity sector, as well as by key innovations. First, the volume of electricity production must be boosted to meet demand. Between 1990 and 2016, it was multiplied by 2.5, from 316.1 TWh to 801.1 TWh. Population growth, urbanization and industrialization on the continent are likely to push those needs even higher, however, increasing electricity demand by 8.9% per year on average between 2015 and 2040.

The necessary production capacity to meet that demand will then be six times higher than that available today. Renewable energy may be a solution (see pp. 98–99). However, innovation is required to serve the most remote populations more effectively. New standalone electricity production systems that do not require expensive connections to the main grid, for example, have already allowed considerable progress and are set to be developed even further.

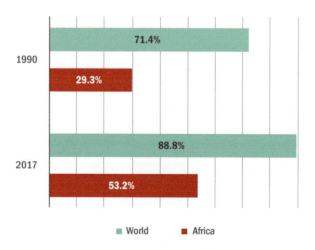

Increase in access to electricity: the world compared to Africa

Source: World Bank (World Development Indicators).

Gourcy, 16 October 2017. The Zagtouli photovoltaic solar power plant in Burkina Faso was inaugurated in 2017 and has an installed capacity of 33 MWh. It has thus boosted electricity production in the country with its estimated annual production of 56 GWh, which represents 5%–6% of the country's annual electricity demand. The commissioning of the power plant and its connection to the national grid through the Zagtouli-Ouahlgouya interconnection project has provided the districts of the city of Gourcy with power connections and street lighting. It has changed the lives of people in this district.
Photo © Erwan Rogard/AFD.

Improvement in access to electricity

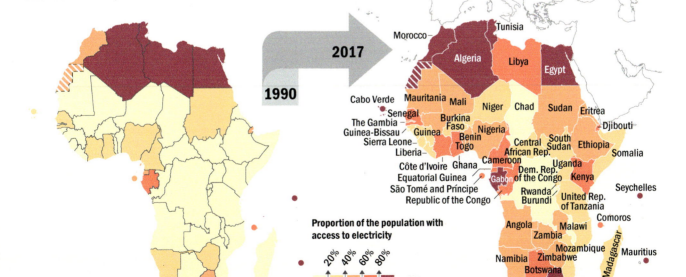

Proportion of the population with access to electricity

20% 40% 60% 80% ND

Source: World Bank (World Development Indicators).

The success of mobile telephones facilitates inclusive banking

THE IMMENSE SUCCESS OF MOBILE TELEPHONES

Whereas mobile telephones have gradually replaced landlines in the rest of the world, in Africa they have imposed themselves from the beginning as the main communication technology. Their success has been astonishing: between 2008 and 2018, the number of cellular subscriptions jumped by 180% on the continent, corresponding to the second strongest growth in the world behind South Asia, where the increase was 213% during the same period. For every 100 inhabitants, there are an average of 83 active subscriptions.

The technological leap has been such that it has made landlines obsolete even before they could be installed, except on an embryonic scale in North Africa. Mobile technology is better suited to the African context, as this digital solution resolves many of the material constraints that were hindering the roll-out of landline networks, such as the lack of linear infrastructures and the changing and informal nature of many settlement areas.

THE MOBILE ACCOUNT, A TOOL FOR ACCESS TO BANKING ON A MASSIVE SCALE

The increased usage of mobile telephones has also revolutionized access to banking services. The development of traditional banking has been hindered by low income levels and limited penetration of the territory by banking networks, with an average of five branches per 100,000 people, against a figure of 10 per 100,000 in South Asia, for example. The creation of mobile payment accounts—similar to bank accounts—has totally reshuffled the cards in financial inclusion.

These personal accounts issued by telecommunications operators with the assent of the banking regulation authorities offer the possibility of converting cash (including very small sums) into electronic currency.

This digital technology appeared in Kenya at the end of the 2000s and has spread widely across the continent, where the number of users (395 million in sub-Saharan Africa in 2018) is 12 times higher on average than elsewhere in the world. This technological revolution has driven a rapid progression in the share of the population that enjoys access to a bank account, up from 17% in 2011 to 38.3% in 2017.

FROM MONEY TRANSFERS TO BANKING SERVICES: HOW MOBILE BANKING IS STRUCTURED IN AFRICA

Africa has a long way to go in developing the possible uses of this new banking technology. Initially, it was used to make payments and transfers from one account to another, in return for payment of a few cents. The technology makes it possible to keep proof of the payments that have been made, thus making them secure. Urban migrants see it as a simple way of sending remittances to members of their families who have stayed behind in their villages, without having to have cash transported over hundreds of miles.

The financial services on offer for the holders of a mobile account are now diversifying rapidly, in the form of more complex products, such as international transfers, microloans or insurance. And governments are beginning to take an interest in the possible uses of mobile banking, as a means of collecting taxes or to reduce the risk of corruption. In this area as in others, it is Africa that is inspiring the rest of the world today.

▌ **Progress in access to banking services**

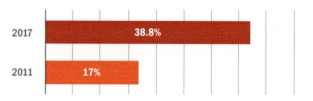

Source: World Bank (World Development Indicators).

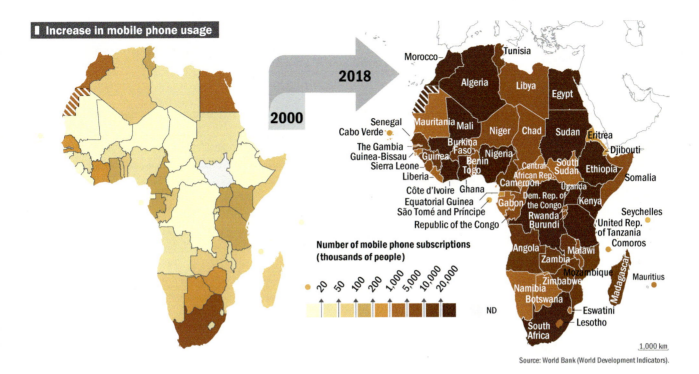

Increase in mobile phone usage

Source: World Bank (World Development Indicators).

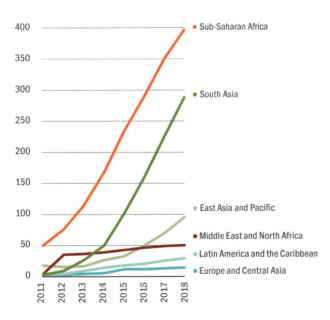

Increase in the number of mobile payment accounts

Source: Global Mobile Money Dataset.

A young woman selling tomatoes at a local market is paid by transfer to her mobile phone

Photo © i_am_zews/Shutterstock.

African economic growth despite instability

CONTRASTING PATHS IN AFRICAN GROWTH

There was a genuine economic turnaround in Africa from the mid-1990s onwards: after 15 years of crisis resulting in a general rise in poverty, the continent has since recorded a marked increase in its wealth. GDP per capita thus rose by 40% between 1995 and 2018 (see graph).

Admittedly, the scale of this increase was less than that in the developing and emerging countries of Asia, which saw their per capita GDP grow almost threefold during the same period, on average. Africa is still home to many of the poorest countries on the planet (see pp. 46–47), and GDP for the continent as a whole represented 2.7% of global GDP in 2018, which is a little less than that of India (3.2%) or France (3.3%). However, there is a striking contrast between the current situation and the continuous weakening of the economy that commenced at the end of the 1980s.

AFRICAN LIONS FIGHTING FOR THEIR DEVELOPMENT

To illustrate these trends, since 2015, 15 countries have recorded average annual growth in excess of 5%, and some of them even rank among the world's 15 countries showing the strongest growth during this period, such as Ethiopia, Côte d'Ivoire, Rwanda, Guinea and Senegal.

Although these economic success stories should not cause us to forget the more difficult situations elsewhere on the continent, they do underscore the diverse trajectories to be found in Africa and show that growth has been maintained at a high level in a certain number of economic powerhouses. Above all, they show that growth drivers have changed in favour of the most diversified countries and to the detriment of those countries that are dependent on commodities (see pp. 44–45).

NEW RESILIENCE IN THE FACE OF THE COVID-19 CRISIS

The economic and social resilience of Africa's economies—the ability of economies and societies to cope with the consequences of unexpected events—is therefore one of the major challenges for the continent. Efforts have already been made and although the rise in commodity prices during the 2000s is one of the reasons behind the strong economic performances recorded during the period (see pp. 42–43), it is no doubt not the only one. Since the end of the economic crisis of the 1990s, the continent has thus benefited from many debt reduction programmes that have stimulated investment in infrastructures, and it has also made considerable economic reforms. There can be no doubt, however, that the progress of recent years requires further consolidation in order to ensure that growth brings a sharper reduction in poverty (see pp. 46–47) The Covid-19 crisis has also shown that in order to achieve more resilient growth on the continent, a reinforcement of welfare systems to make them more effective and increased economic diversification are necessary, among other things.

Wealth per capita in Africa

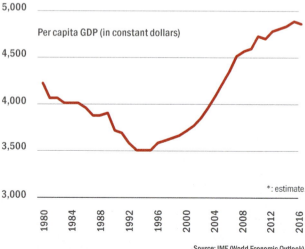

Source: IMF (World Economic Outlook).

A man from Mopti in Mali. He has been a vegetable gardener for more than 10 years. Thanks to the ACTIF project (Support for Youth and Local Authorities in their Vocational Training and Integration Initiatives), he has been able to make investments and increase his **production yields.** Photo © Harandane Dicko (May 2019).

■ The world's fastest-growing countries (2015–2019)

Average annual rate of growth in real GDP during the period 2015–2019.

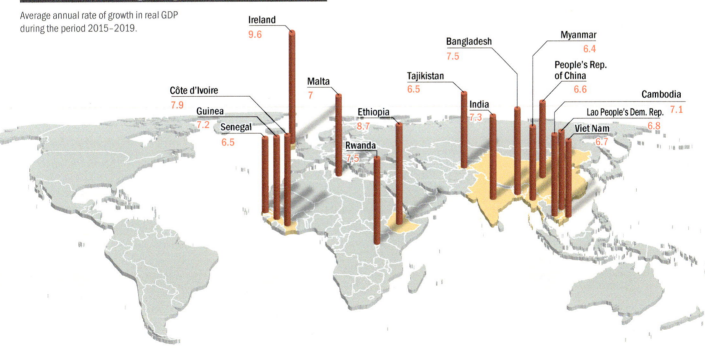

Ireland 9.6
Malta 7
Côte d'Ivoire 7.9
Guinea 7.2
Senegal 6.5
Ethiopia 8.7
Rwanda 7.5
Tajikistan 6.5
Bangladesh 7.5
India 7.3
Myanmar 6.4
People's Rep. of China 6.6
Cambodia 7.1
Lao People's Dem. Rep. 6.8
Viet Nam 6.7

Countries enjoying strong growth

Source: IMF (World Economic Outlook).

Africa's asset: the internal market

AN ECONOMY ON A PATH OF SUSTAINED GROWTH

The size of the African economy, and therefore of its internal market, is one of the continent's assets: in 2020, it represented $2.6 billion, which is roughly equivalent to the French economy in the same year. Strong population growth and the rise in GDP per capita are also driving the expansion of the economy. For instance, if Africa continues to grow at the same rate as it has over the past 20 years, i.e. at an annual rate of 5.5%, in 2050 the size of the African economy will be 75% greater than that of France. Above all, this is an asset for Africa itself, where the internal market is likely to provide outlets for its own output, notably from agriculture.

TOWARDS A CONSUMER MARKET?

Does that mean that we are seeing the emergence of a middle class capable of making not just essential expenditure (food, housing and healthcare), but also of spending that is less directly vital (education, vehicles or telephones, for example)?

In a report published in 2011, the African Development Bank stated that this was indeed the case, estimating that the middle class on the continent—corresponding to that part of the population earning between $4 and $20 per day—had doubled between 1980 and 2010, from 60 million to 123 million. According to more recent data, this middle class is thought to represent 14% of the population at present, to which should also be added the 6% who earn more than $20 per day. Their household purchasing power remains limited, however, as 64% of these middle classes earn less than $10 per day.

The middle class is also characterized by criteria other than income, such as social status, the type of job or other cultural norms. Various studies have shown the heterogeneous nature of Africa's middle classes, composed of informal sector workers and senior executives from the private sector, farmers and also civil servants in managerial positions. Their needs and spending capacities may therefore vary widely and the strategies for responding to those needs must be different. For producers, the challenge is therefore to develop products and services that are tailored to the specific socio-cultural features and purchasing power of local consumers. Leading Western brands and also many local or up-and-coming brands are already hard at work trying to do so.

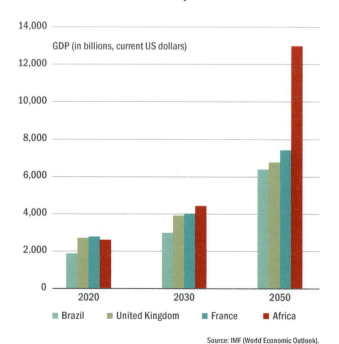

Growth of the African economy
GDP (in billions, current US dollars)
Brazil, United Kingdom, France, Africa
Source: IMF (World Economic Outlook).

Panoramic view of Jemaa el-Fnaa Square and the market of the Medina in Marrakesh, Morocco. Photo © Antonio Rico/Shutterstock.

ESTABLISHING AN INTERNAL MARKET

The African market still tends to be made up, however, of a number of different centres of consumption, scattered across the continent and disconnected from each other by natural, political and commercial borders. However, a process of concentration can be observed around regional consumption centres in which urbanization processes are shaping internal markets that are relatively attractive.

This is the case in the largest emerging economies, such as Nigeria, South Africa, Egypt and the Maghreb countries, and also countries such as Ethiopia, Kenya and Côte d'Ivoire. The creation of the African Continental Free Trade Area (AfCFTA) to connect the markets with one another and thus to facilitate trade constitutes another challenge to enable Africa to profit fully from its internal market (see pp. 58–59 and 60–61).

Distribution of daily income among the African population (2015)

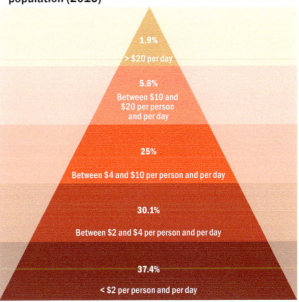

- 1.9% — > $20 per day
- 5.6% — Between $10 and $20 per person and per day
- 25% — Between $4 and $10 per person and per day
- 30.1% — Between $2 and $4 per person and per day
- 37.4% — < $2 per person and per day

Source: World Bank (Povcalnet).

A changing economy pursuing its own development path

AGRICULTURE: A FAST-CHANGING PIVOT OF THE ECONOMY

The African economy is still based on agriculture: the farming sector continues to employ almost half of the continent's working population. However, its role is changing.

First, the share of agriculture in employment has dropped by 6 percentage points over the past 20 years. We should not be misled by this figure, because the number of people employed in the agricultural sector is still growing, driven by the rapid increase in the population. However, of the total number of people entering the labour market, only a small share is going into this sector.

Second, agriculture now represents a minority share in wealth production—about 17% of African GDP. This gap between the importance of the sector in terms of employment and its limitations as a means of wealth production is a reflection of limited productivity levels (see pp. 112–113).

The share of farming in African GDP has levelled out over the past 20 years, thus, testifying to the modernization efforts that have been made.

THE AFRICAN CONTINENT: A SERVICE ECONOMY

So apart from the farming sector, where do Africans find jobs?

Mainly in services. These include, among others, banking and finance, the transport sector, commerce or the state civil service. Although services have long been the main area of activity in Africa, the sector has expanded recently: the share of services-related employment out of the total number of jobs in Africa rose from 33% to 38% between 2000 and 2018, while the share of African wealth produced by the sector leapt from 50% to 56% in less than 20 years.

While these jobs are often in informal activities in commerce, transport and catering, the figures do show a clear underlying trend: that of the strong performance of this sector is of key importance to the African economies, albeit that it is sometimes underestimated and misunderstood.

INDUSTRIALIZING AFRICA: A PATH YET TO BE FOUND

And what about Africa's industrial potential? As things stand today, the results are not satisfactory and the share of wealth produced by this sector is declining (see pp. 114–115). There are two reasons for this trend.

The first is a weakening of the extractive sector—made up of the mining and petroleum industries—in recent times, driven by the fall in commodity prices: in 2017, this sector represented under 15% of the wealth produced by the continent. We are a long way here from the stereotype of African economies as being based on the exploitation of resources, although these do still continue to make up a large share of exports and state income for a large number of countries (see. pp. 44–45).

The second reason for this decline in industry is the fall in the share of the manufacturing sector, comprising transformation and assembly activities among others, in the continent's GDP. This sector, which has played a key role in the economic development of countries in Asia and Europe, is struggling in Africa. The development of industry is being held back by a shortage of infrastructures (e.g. transport, telecommunications and energy), which breaks up value chains, a lack of regional integration and low human capital. Although the importance of the sector does vary from one country to another (see map), the predominant model in Africa is still marked by a lack of transformation of commodities, which are often exported unprocessed (see pp. 114–115).

Closer economic ties with the rest of the world

AFRICA'S INTERNATIONAL TRADE TRANSFORMED BY THE EMERGING COUNTRIES

Over the past 20 years, trade between Africa and the rest of the world has been transformed by the rise of the emerging countries, and in particular of China. The latter's share of African exports thus leapt from 3% to 14% between 2000 and 2018. Similarly, while China supplied only 3% of the goods purchased by African countries outside their own borders in 2000, this share had risen to 16% by 2018.

The proportion of trade with Africa's traditional partners—the so-called developed countries—has fallen during the same period: they now purchase 45% of Africa's exports, compared to 71% in 2000, and supply Africa with 40% of its imports, compared to 63% in 2000. Another key trend is the development of intra-African trade, which now represents 14.5% of total trade in Africa (see pp. 60-61).

ECONOMIC EXCHANGES BETWEEN AFRICA AND THE REST OF THE WORLD HAVE GROWN SHARPLY

This decline in the share of Africa's traditional partners does not mean, however, that they are doing less trade together. On the contrary, it is the world as a whole that is doing more trade with Africa: African trade amounted to $928 billion in 2018, compared to $235 billion in 2000. Trade between the European Union and Africa thus grew from $98 billion in 2000 to $266 in 2018.

This increase in African trade by value echoes that in investment: in the space of just 20 years, the volumes of investment by foreign countries in Africa have grown fivefold, leaping from $9.7 billion in 2000 to $45.9 billion in 2018.

AFRICA, A SMALL PLAYER IN INTERNATIONAL TRADE, BUT CLOSELY CONNECTED WITH THE REST OF THE WORLD

Despite the improvement in the continent's exports and imports of goods, they represent only a small share of world trade (2.4% in 2018). And yet Africa is

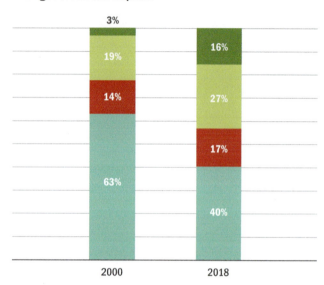

Origin of African imports

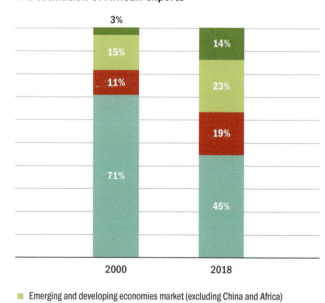

Destination of African exports

- Emerging and developing economies market (excluding China and Africa)
- Developed countries market
- Intra-African market
- China

Source: IMF (Direction of Trade Statistics).

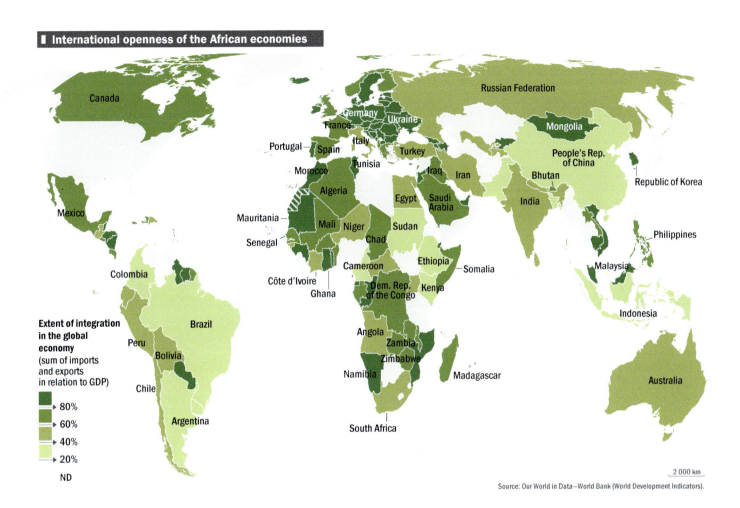

one of the regions that is the most reliant on international trade. The value of trade in goods, meaning exports and imports, thus represents half of the size of the economy for the vast majority of African countries.

Africa must import a significant share of the goods it consumes (foodstuffs, machines and capital goods, pharmaceutical products, etc.) For some countries, and in particular those that produce oil or minerals, exports also represent an essential source of revenues (see pp. 44–45).

This marks a sharp contrast with China, for example, a central player in world trade, where trade in goods represents just 38% of the economy.

Africa, the new travel destination

THE INCREASE IN AIR TRANSPORT

Aircraft are landing and taking off in growing numbers in Africa. Air traffic has been growing constantly for 20 years now: the number of passengers has increased from 31.6 million travellers in 2000 to 95.2 million in 2018. Although the continent represents just 2.2% of worldwide passenger transport by air, growth in traffic to and from Africa (6.2%) has exceeded the world average (5.4%) (see graph). This trend has fostered the emergence and consolidation of an African air transport industry, composed of airport infrastructures and airline companies. In 2019, 35 countries had a national airline and 13 of them had at least two airline companies registered locally.

While this development provides better connections between Africa and the rest of the world, it has also provided a boost for intra-African routes, which now represent 27% of air traffic in Africa. The increase in international traffic has been of particular benefit to the airlines of certain countries that have succeeded in tapping into this emerging demand, such as Ethiopia, Egypt, South Africa and Morocco.

The Namib Desert and Skeleton Coast from the air, Namibia. Photo © Cezary Wojtkowski/Shutterstock.

RAPID GROWTH IN PASSENGER FLOWS TO AFRICA AND WITHIN THE CONTINENT

Growth in air transport has no doubt boosted the growth in tourist flows on the continent: almost one in two tourists arrives on the continent by plane. Since the beginning of the 2000s, the number of international arrivals in African countries has more than doubled, from 32.1 million in 2000 to 70.1 million in 2017 (see map).

Another key tendency has been that Africa is also a domestic source of tourists: over 40% of tourists came from another African country in 2016. Europe, meanwhile, continues to have a special relationship with Africa and provides some 45% of non-African tourists.

TOURISM POTENTIAL TO BE EXPLORED

Although the number of tourists is increasing constantly in Africa, the continent is not yet drawing all the potential benefits from this. Income from international tourism increased only modestly, from $17 billion to $50 billion between 2000 and 2008, and has remained at that level since then (see graph). In reality, only 15% of travellers go to Africa to take holidays there. The remainder include businessmen and women (7.3%), as well as people travelling for personal reasons, and in particular members of the diaspora returning to visit their country of origin. The latter are usually accommodated by their family in the country and therefore rely less on the tourism industry.

However, the development of tourism remains an opportunity for a large number of African countries. At present, four countries capture over half of the tourist numbers and revenues on the continent: South Africa, Egypt, Morocco and Tunisia. But other countries could no doubt promote their historic and natural heritage sustainably for the greater benefit of their populations.

The increase of tourism in Africa

Number of tourists: 200,000 – 500,000 – 1,000,000 – 1,500,000 – 6,000,000 – ND

Source: World Bank (World Development Indicators).

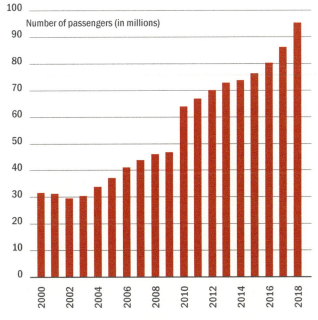

Expansion in air traffic in Africa

Number of passengers (in millions)

Source: World Bank (World Development Indicators).

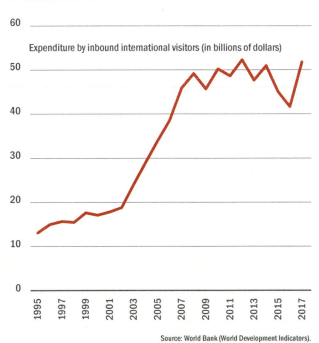

Trends in tourism revenues

Expenditure by inbound international visitors (in billions of dollars)

Source: World Bank (World Development Indicators).

The courting of Africa at the diplomatic and cultural level

AFRICA RESTORED TO PROMINENCE IN DIPLOMACY

The interest in Africa is not just economic, and can also be seen at the diplomatic level. A large number of embassies have been inaugurated in recent years on the continent—over 320 since 2010. Most of the major Northern countries are now represented in all the countries of Africa (see map), as are several emerging countries: for instance, the network of Turkish embassies in Africa has quadrupled since 2005, from nine in 2003 to 41 in 2015, while Brazil now has one of the largest networks of embassies in Africa.

The number of high-level presidential and governmental visits to the continent has also increased since the end of the 2000s, the heads of state of the major powers (China, the United States, Japan, the Russian Federation, the European Union) and of the emerging nations (Brazil, Turkey) have been making regular visits there.

"AFRICA" SUMMITS AND CULTURAL INSTITUTES.

These increasingly close relationships with other countries can also be seen in the growing number of events bringing the continent together with its partners or the opening of cultural institutes.

For example, five Africa-European Union summits have been held since 2000 and France has organized 28 France-Africa summits since 1973. Japan's TICAD (Tokyo International Conference on African Development) has been held every three years since 2016 compared to every five years previously. Since its first edition in Beijing in 2000, China has organized the Forum on China–Africa Cooperation (FOCAC) every three years. India, meanwhile, organized an India-Africa summit in 2015, while the Russian Federation held its first Russia-Africa summit in 2019.

The presence of cultural cooperation agencies on the continent is another way of establishing closer relations with Africa. There are *Alliances Françaises* and *Instituts Français*, British Councils and Goethe Institutes promoting French, British and German culture established in 46, 25 and 26 countries, respectively, while China possesses a network of 61 Confucius Institutes on the continent, in 46 countries.

AFRICA AT THE UNITED NATIONS

If Africa's voice counts so much, it is notably because of its place within the United Nations (UN). With its 54 recognized countries, the continent represents more than one-fourth of the organization's member states.

Gathered together in the Africa regional group—originally along with the Asian countries, but on their own since 1963—the African countries with one another to bring their weight to bear on negotiations, debates and votes. The Africa Group also has three non-permanent seats at present (out of a total of 10 available) on the UN Security Council, to be shared with the Asia Group every two years and for two years. For the moment, 45 of the 54 African countries have been non-permanent Council members, and the continent intends to increase its influence there.

In 2005, the African Union adopted a common African position on reform of the UN, the Ezulwini Consensus, which involves reserving at least two permanent and five non-permanent seats for the continent.

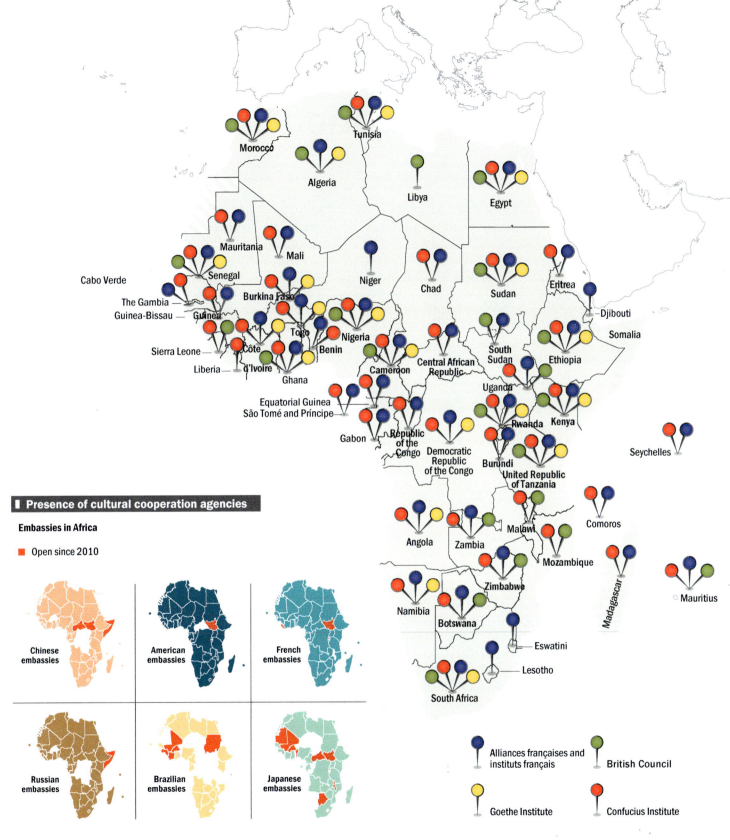

Political liberalization in Africa

THE IRRESISTIBLE SHIFT TOWARDS DEMOCRACY ON THE CONTINENT

From Egypt to Tunisia, Burkina Faso to Gambia and Sudan to Algeria, Africa has seen a remarkable number of political upheavals in recent years, driven by the grass-roots population. These events are all part of an underlying trend towards democratization on the continent.

Over the past 50 years, at no time has Africa had so many political systems born out of multi-party elections. Back in 1985, over 90% of the countries were governed by autocratic regimes, but by 2015 almost half of the continent had more democratic regimes (see map).

However, in many countries, the regimes remain hybrid. Although the organization of multi-party elections fulfils the formal criteria for a government to be considered representative, major problems do still persist in the areas of election integrity, working conditions for the media and opposition parties, and the application of checks and balances.

STRONG GRASS-ROOTS ASPIRATIONS FOR GREATER DEMOCRACY

The progress achieved by democracy as a political regime varies from one region to another, with Southern Africa and the Gulf of Guinea being those where it would appear to be establishing itself most firmly. Such transformation does constitute an aspiration that is shared throughout the continent, however. The population of Africa holds strongly to the democratic ideal and the proper functioning of the institutions. The figure of the single party or authoritarian leader, whether elected or military, is now poorly perceived in society, while support for the rule of law is almost unanimous.

In such conditions, surveys of the population reveal that demand for democracy far outstrips supply of democracy as currently perceived by the population, giving rise to strong grass-roots calls to that effect all over the continent (see graph).

THE RULE OF LAW: A GAP BETWEEN THE INSTITUTIONS AND PRACTICES

The progress achieved in matters of democracy does still hide some darker realities. In particular, popular discontent is focused on the gap between the institutions and practices, especially in matters of breaches of human rights, freedom of the press and corruption.

The continent has made significant advances, however, and these weaknesses are part of the normal democratization process, the duration of which may vary according to the specific features of each country.

■ Demand for and supply of democracy in Africa

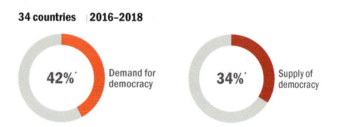

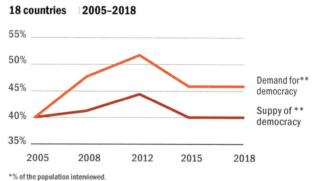

*% of the population interviewed.
**% supply and demand perceived by African citizens according to surveys.

Source: Afrobaromètre.

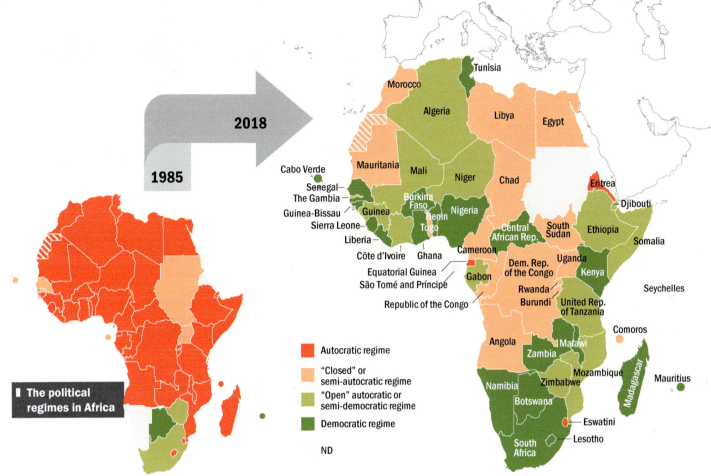

The political regimes in Africa

- Autocratic regime
- "Closed" or semi-autocratic regime
- "Open" autocratic or semi-democratic regime
- Democratic regime
- ND

Source: Center for Systemic Peace (base Polity IV)

The Polity IV indicator comprises five variables measuring: (i) the institutional constraints on the decision-making powers of the chief executive; (ii) the degree of institutionalization, competition and transparency of the political leader selection mechanisms; (iii) the degree of openness of the executive recruitment process; (iv) the degree of institutionalization or regulation of political competition; (v) the extent of government restriction of political competition.

Counting votes in a polling station in Gabu, Guinea-Bissau, during the 2014 general elections.

Photo © aborbasch/Shutterstock.

Less lethal conflicts, more complex violence

A REDUCTION IN THE INTENSITY OF ARMED CONFLICTS BETWEEN STATES

Although Africa has suffered lethal conflicts, and without seeking to deny the security risks in certain regions, the number of armed conflicts has been on a downwards trend since the 1990s. All the regions of Africa have known devastating wars in the past, whether in Algeria, Sierra Leone, Somalia or the Democratic Republic of the Congo.

However, as has been seen elsewhere in the world, the number of armed conflicts in Africa involving a state among the belligerents has fallen drastically since the end of the Cold War. Not only have wars been fewer in number, they have also been less lethal. The number of victims in conflicts involving a state has thus been on a downward trend since the mid-1980s, falling from 53,307 to 9,342.

THE CHANGING FACES OF VIOLENCE

As the intensity of conflicts between states lessens, the nature of that violence is changing and becoming more complex, with the involvement of new actors (rebel and extremist groups). This trend concerns two main forms of violence.

First, the activities of terrorists and armed groups have developed since the 2000s. Non-state conflicts involving actors other than states have almost doubled since 2010, with the number of this type of conflict increasing to 284 during the period 2011–2018, according to the Peace Research Institute Oslo (PRIO) in Oslo. Second, political violence (riots, violence against the civilian population, etc.) has grown, especially in urban areas, driven by the rise in political demands.

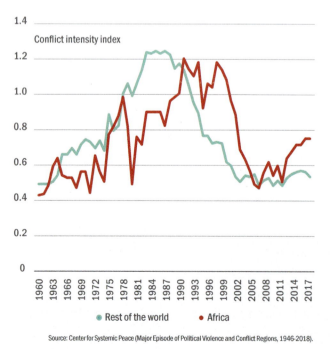

Conflicts on a downward trend

Source: Center for Systemic Peace (Major Episode of Political Violence and Conflict Regions, 1946-2018).

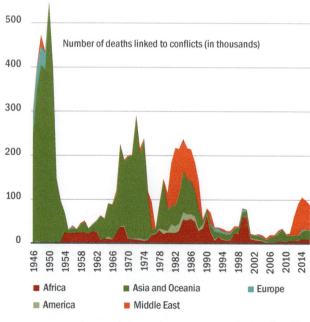

Less lethal state conflicts

Source: Our World in Data (UCDP/PRIO).

GEOGRAPHICALLY LIMITED CONFLICTS

These armed conflicts do not concern the whole of Africa and today remain concentrated in a few countries. During the period 2011–2017 period, for instance, the major conflicts representing most of the victims were concentrated in seven countries: Sudan, Nigeria, the Democratic Republic of the Congo, Somalia, South Sudan, the Central African Republic and Libya (see graph). More recently, the situation has also deteriorated rapidly in the Sahel.

Conflicts are thus spreading along two main axes, with the first running from Mali to Somalia via Nigeria, and the second from Libya to Congo via Sudan. This distribution corresponds to that of those countries said to be the most fragile, characterized by the weak capacity and/or legitimacy of the state (see pp. 48–49).

In addition, conflicts have hitherto concerned mainly those parts of national territories that are the furthest from the centres of power and the economy.

The deterioration in the security situation is often the result of a series of factors: weak control over the territory by the state; a feeling of exclusion among a part of the population; and/or unsatisfied expectations in matters of development. Consolidating peace constitutes a major challenge for the development of the continent, and thus requires security, political and also development responses.

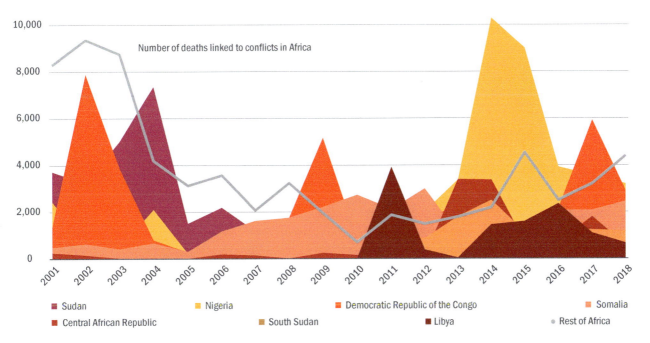

The most lethal conflicts concentrated in a few countries

Source: UCDP (Georeferenced Event Dataset).

The gradual improvement in economic and public governance

INCREASING GROWTH FOSTERED BY IMPROVEMENTS IN THE MACRO-ECONOMIC FRAMEWORK

The period of economic expansion enjoyed by Africa since 2000 (see pp. 20–21) is often explained by the rise in commodity prices, but there can be no doubt that it is also linked to some extent to a transformation in the macro-economic framework, which is more difficult to measure. The debt crisis of the 1990s and the resulting debt relief process—the Heavily Indebted Poor Countries (HIPC) initiative—went hand in hand with structural reforms of the economy targeted at resolving some of the main sources of macro-economic imbalances. Profound changes were made to all the different dimensions of countries' economic policies—monetary, budget, finances, regulatory and state participation in the economy.

These changes in macro-economic management proceeded parallel to those that were taking place in political governance (see pp. 32–33) and thus enhanced the management capacity of government administrations. Improved macro-economic management in all the African countries inspired confidence in investors and consumers alike, and thus has remained one of the decisive drivers of this growth.

A MORE CONDUCIVE ENVIRONMENT FOR COMPANIES, BUT WITH CONSIDERABLE VARIATIONS

One of the most obvious areas in which these transformations have been visible is that of reforms to foster private initiatives. According to the World Bank's Doing Business index, which measures ease of doing business, Africa has been the continent which has made the most regulatory changes to improve the business environment each year since 2013.

In 2019, 89 reforms were completed in African countries, representing 30% of the reforms made all over the world that year. Over the past decade, several countries on the continent have featured among the world's best reformers. In 2019, for example, five of the 15 countries that made the most significant improvements were African (Senegal, Togo, Nigeria, Niger and Zimbabwe). However, although these improvements to the business climate are a general trend on the continent, there are still big differences from one country to another. While Mauritius (ranked 13th) and Rwanda (ranked 38th) are now among the highest performing economies, 28 African countries are in the bottom fourth of the ranking, with the fragile states ranked lowest.

ENCOURAGING GAINS AND REFORMS TO BE CONTINUED

However, public governance cannot merely be reduced to a question of macro-economic governance and business environment. According to the Mo Ibrahim Foundation, it concerns "the provision of the political, social and economic goods that a citizen has the right to expect from his or her state, and that a state has the responsibility to deliver to its citizens". According to the index compiled by the association, overall governance in Africa has maintained its improving trend. In 2017, the continent posted its highest score in the past 10 years (2008–2017). Thirty-four countries have improved their governance over the past 10 years.

However, this progress is being driven by a small group (made up in particular of Côte d'Ivoire, Zimbabwe and Morocco) and the gaps between countries are widening. While the scores of all the countries on the continent were close to the African average during the first few years of the decade, the gaps have been widening over the past 10 years. In 2017, nine countries posted their worst performance of the decade in overall governance, while 19 others achieved their best performance.

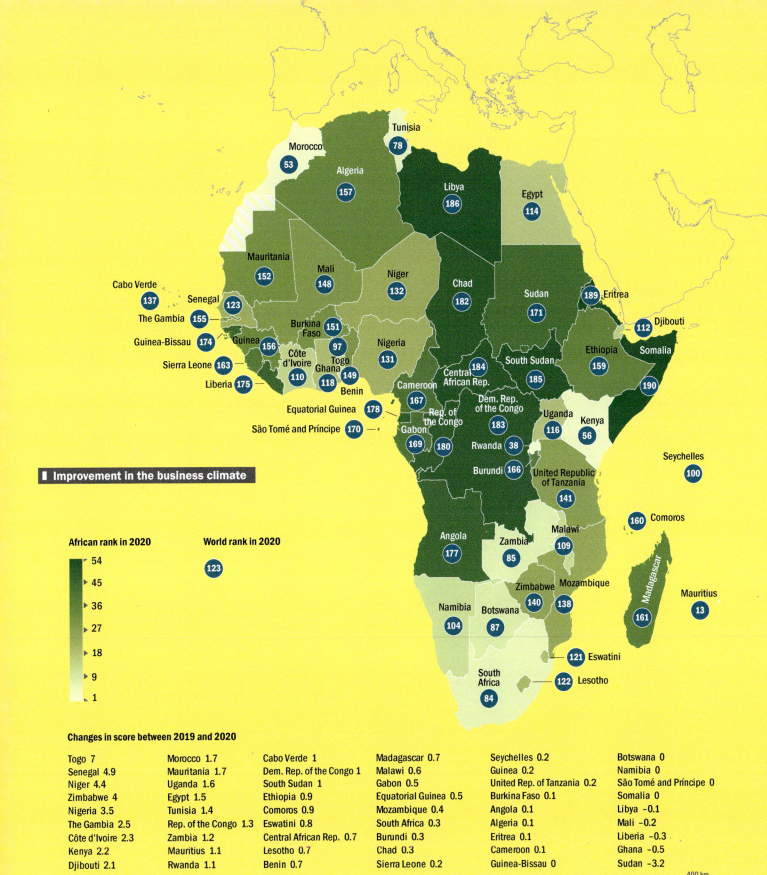

A MULTIFACETED CONTINENT WITH SHARED CHALLENGES

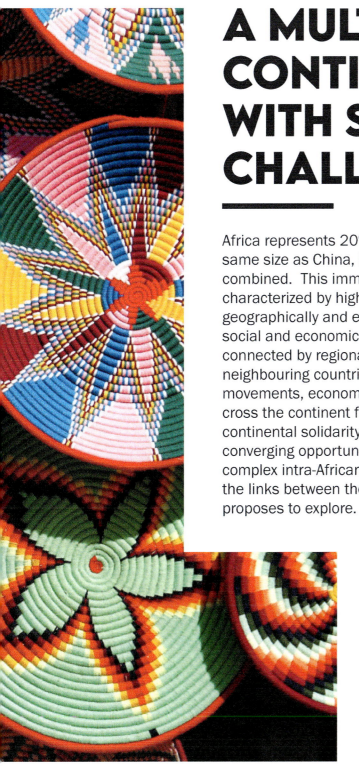

Africa represents 20% of the world's land area, making it the same size as China, India, the United States and Europe combined. This immense continent is far from uniform and is characterized by highly diverse situations and dynamics, not only geographically and environmentally, but also at the cultural, social and economic level. Its borders notwithstanding, Africa is connected by regionalization processes that are bringing neighbouring countries closer together politically, and by human movements, economic exchanges and natural systems that cross the continent from north to south and east to west. This continental solidarity forms a space for shared challenges and converging opportunities, giving rise to increasingly solid and complex intra-African relations. It is the many faces of Africa and the links between them that the second part of this Atlas proposes to explore.

Traditional Ethiopian baskets.
Photo © Evgenii Zotov/Getty Images.

African giants and small economies: how to read the African economy

TEN MAJOR ECONOMIES STAND OUT ON THE CONTINENT

Ten African countries stand out owing to the size of their economies: Nigeria, South Africa, Egypt, Algeria, Angola, Morocco, Kenya, Ethiopia, Ghana and the United Republic of Tanzania. Together, they represent almost three-fourths of the total wealth produced on the continent as a whole.

In addition to the size of their GDP, these countries also stand out owing to their demographic weight. Nigeria, Ethiopia and Egypt alone—which have between 100 and 200 million inhabitants each—represent almost one-third of the population of the continent as a whole. Other countries with more modestly sized economies also count among those with the biggest populations in Africa, such as the Democratic Republic of the Congo, Uganda and Sudan (about 42 million inhabitants each).

A DECISIVE ROLE IN THE OVERALL PERFORMANCE OF THE CONTINENT

On account of their economic weight within the continent, the economic health of these countries largely shapes the way the overall trajectory of Africa is perceived. The aggregate figures for African growth are therefore largely tied to the situation in a few of these countries. The slowdown observed in Nigeria, Algeria and Angola after the collapse in oil prices in 2014, or weak growth in South Africa since the financial crisis of 2009, therefore go a long way towards explaining the weaker aggregate performances of the continent in recent years.

There can be no doubt that we should not seek to downplay the importance of these trends: on account of the size of their populations, what happens in these countries economically affects many people living on the continent. Their economic situation also influences that of neighbouring countries (as is the case for South Africa in the southern part of the continent, or Nigeria for the countries in West Africa). However, this can hide other dynamics at play in other African regions and countries.

THE LARGEST CONCENTRATION OF THE WORLD'S SMALLEST ECONOMIES

Africa is the continent that concentrates the largest number of small economies. Among the 30 countries in the world with the lowest levels of GDP (not counting island states), 20 are African: their cumulative GDP stood at around $120 billion in 2018, which is roughly equivalent to the GDP of Morocco.

Each of these economies also has less than 26 million inhabitants, which considerably limits the size of their internal market and increases their reliance on foreign countries, to whom they must turn for sourcing and outlets for their production. Integration with the other African economies is therefore a crucial challenge for their development.

Photo © Billion photos/Shutterstock.

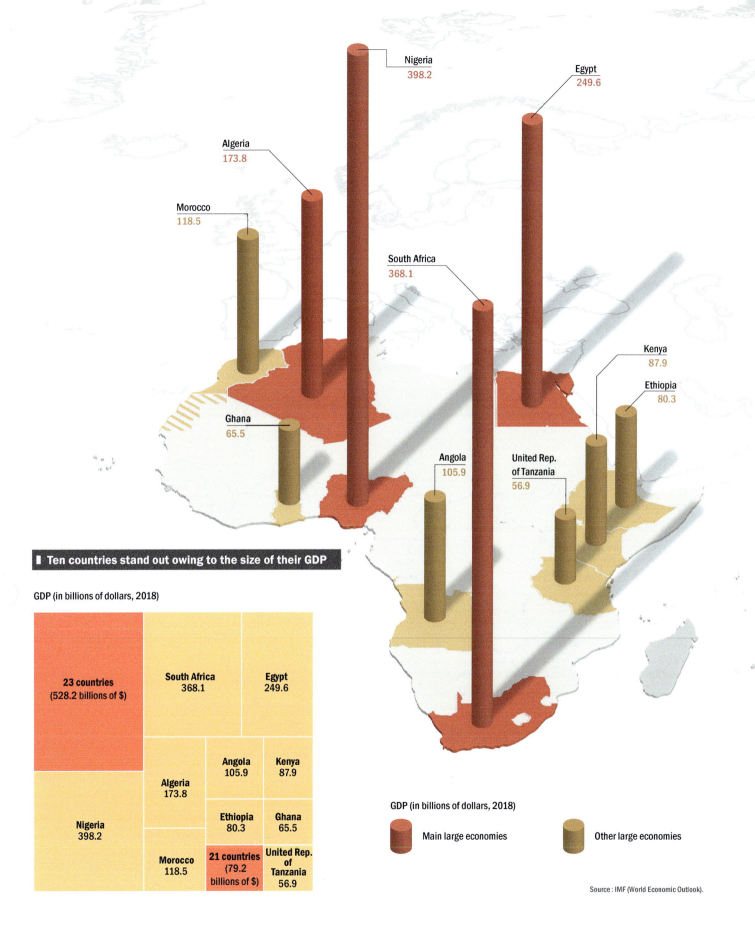

Ten countries stand out owing to the size of their GDP

Source: IMF (World Economic Outlook).

The geography of wealth in Africa

TWO CENTRES OF WEALTH IN THE NORTH AND SOUTH OF AFRICA

The GDP per capita indicator—an indicator of the wealth of countries, unlike GDP which measures the size of their economies—is no doubt less than perfect in many ways. More particularly, it does not reflect the real extent of poverty and inequalities prevailing within a country (see pp. 46–47 and p. 50–51). It does, however, reveal a sometimes little-known geography of wealth within the continent.

What surprises us when we look at the countries of Africa according to their level of wealth per capita is the polarization between the north and the south of the continent: these two regions are home to half of the 20 countries with the highest GDP per capita on the continent. And yet it is neither in Southern Africa nor in North Africa that we find Africa's four richest countries. The top two in terms of GDP per capita are two Indian Ocean islands, Seychelles and Mauritius, which have built their development more particularly on tourism and banking services, or on textiles. They are followed by two oil-producing countries in Central Africa—Equatorial Guinea and Gabon.

THE POOREST COUNTRIES ARE SCATTERED ALL OVER THE CONTINENT

The 10 poorest countries, meanwhile, are scattered through all the regions of the continent (except for North Africa): in Central Africa (the Central African Republic and the Democratic Republic of Congo), Southern Africa (Mozambique and Malawi), East Africa (Burundi and South Sudan), West Africa (Liberia, Sierra Leone and Niger) and in the Indian Ocean (Madagascar).

Looking at this distribution, we may see that being landlocked geographically seems to be a major drawback: except in Southern Africa, no African country without access to the sea has a GDP per capita in excess of $1,000.

Nevertheless, while geographical location seems to hand out opportunities and challenges, it is not the only factor driving poverty.

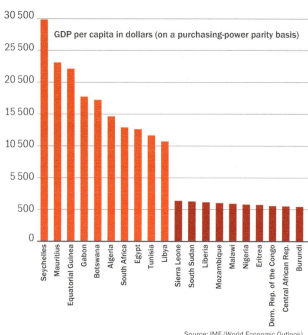

Extremes in terms of wealth (2018)

Source: IMF (World Economic Outlook).

TRENDS AS PERCEIVED THROUGH THE INTERNATIONAL RANKINGS

On a worldwide scale, the international rankings show that Africa continues to concentrate a large portion of the world's poorest countries. It is in Africa that 80% of the low-income countries are to be found, according to the World Bank, meaning those with a GDP per capita of less than $1,025 in 2018. The continent is also home to 70% of the least developed countries (LDCs), which are considered to be most vulnerable in the world.

A growing number of countries are now entering the category of so-called middle-income countries, i.e. those whose per capita income was above the threshold of $1,025 in 2018. Some of them, like Kenya or Senegal, have entered this category just recently, while other fast-growing countries, such as Morocco, Ghana and Côte d'Ivoire, had already passed the threshold and are seeing their GDP per capita continue to progress.

View of tower blocks in Port Louis, capital of Mauritius. Photo © Fabien Dubessay/AFD.

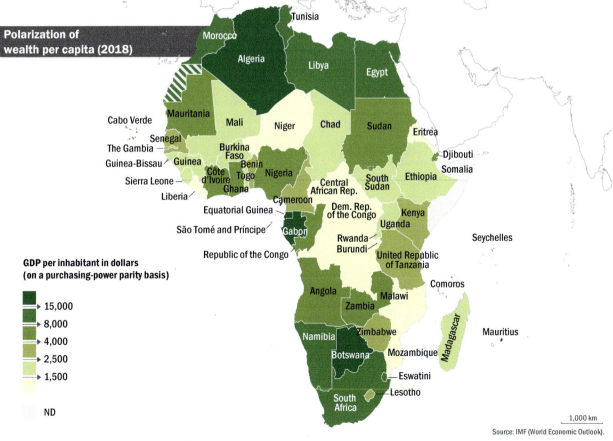

Polarization of wealth per capita (2018)

GDP per inhabitant in dollars (on a purchasing-power parity basis)
- 15,000
- 8,000
- 4,000
- 2,500
- 1,500
- ND

Source: IMF (World Economic Outlook).

Considerable natural resources: who for? What for?

THE WEIGHT OF COMMODITIES IN THE AFRICAN ECONOMIES

Contrary to what is often thought, for the great majority of African countries, crude petroleum, gas and mining resources represent only a limited portion of their economy. The share of such commodities exceeds 20% of GDP in just 10 African countries out of 54, and the figure is barely any higher when one includes agricultural output which is most often intended for export (coffee, cocoa, cotton, etc.).

Does that mean that there is a misunderstanding as to the importance of commodities to economic development on the continent? Probably not: for certain states, export sales of these resources are the main source of currency (sometimes the only source) and they are essential to finance their imports. Income from the exploitation of commodities may also represent a large share of the income of these states, when they have difficulty taxing other economic activities, which are sometimes informal.

LESS THAN HALF OF THE AFRICAN COUNTRIES ARE RICH IN NATURAL RESOURCES

Only 25 countries in Africa are considered "rich in natural resources", to such an extent that they represent a significant share of their exports. But rather than their volume, it is in fact the importance of the economies in question in terms of the size of their GDP that counts: six of the 10 biggest economies (Nigeria, South Africa, Algeria, Angola, Ghana and the United Republic of Tanzania) hold most of the natural resources.

In these conditions, changes in commodity prices have a big effect on the overall trend for the continent. And this is all the more true insofar as in a certain number of countries—Angola, Algeria, Libya, Nigeria, Sudan—exports of natural resources can represent more than 90% of export income.

THE LEAST COMMODITY-RELIANT COUNTRIES ARE THE MOST DYNAMIC

Over the course of the past decade, the fall in prices of oil and other commodities has severely disrupted the economies on the continent that were exporting them. In this way, being rich in natural resources is no longer a source of growth, at least for the moment. On the other hand, it is now those countries that are relatively well diversified (Morocco, Senegal and Mauritius, for example), and therefore the least reliant on natural resources, that are on the most stable paths with the highest growth rates.

What is more problematic is that over-investment in the natural resources sector has often come to the detriment of other activities, with the result that the rest of the economy is struggling to emerge.

This commodities trap, in which many countries around the world have fallen, including in Europe, is not inevitable, however. Botswana, a big diamond producer, has set up a system for good management of its mining resources, in order to reallocate mining revenues to the education and health budgets, and to invest in infrastructures in order to reduce its reliance on this resource. In addition to this, since 2011, Botswana has developed a programme to process and market its diamonds locally. The rough diamonds are now cut in and sold from the country.

Extreme poverty concentrated on the continent

A DECLINE IN POVERTY RATES ACROSS THE CONTINENT

On the whole, Africa has recorded a marked fall in rates of extreme poverty over the past decades. The proportion of people living on less than $1.90 per day thus fell from 47% of the population in 1985 to 39% in 2017 (see graph).

The increase in the number of poor people on the continent as a whole, which may appear to contradict this trend in extreme poverty, can be explained by the level of population growth which is still too high to offset the rate at which poverty is being reduced. It suggests that even if the progress of development is being shared by a growing portion of the population, they are not yet managing to reach the critical expansion threshold to put an end to the growth in the number of people living in extreme poverty, which reached 469 million individuals in 2017 at the level of the continent.

EXTREME POVERTY, A PHENOMENON THAT IS INCREASINGLY SPECIFIC TO AFRICA

Extreme poverty is gradually becoming very much an African singularity. While the continent was home to one-fourth of the world's extreme poor in 1990, the figure is now over 60% of them (469 out of 751 million people), and is likely to reach 90% by 2030 (see graph).

While it is the announced eradication of the phenomenon in East and South Asia that is largely contributing to this relative concentration effect, the number of extremely poor Africans is continuing to increase at the same time, notably in the Sub-Saharan regions. Of the 30 countries on the planet with the highest rates of extreme poverty today, 27 are located in Africa.

Trends in the rate of extreme poverty

Source: World Bank (World Development Indicators).

Projections of the number of poor people

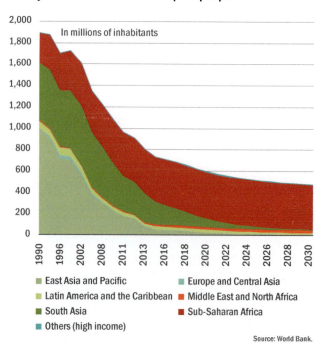

Source: World Bank.

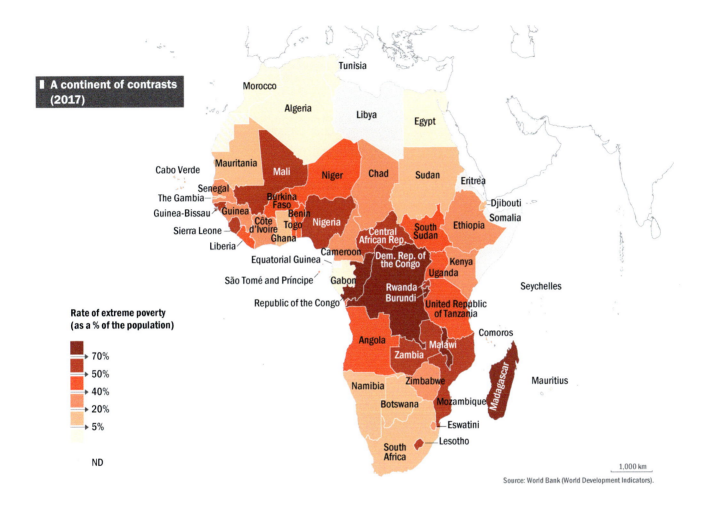

A continent of contrasts (2017)

Rate of extreme poverty (as a % of the population)
- 70%
- 50%
- 40%
- 20%
- 5%
- ND

Source: World Bank (World Development Indicators).

POVERTY CONCENTRATED ON THE CONTINENT

Looking beyond this global overview, progress in reducing extreme poverty is shared out very unevenly on the continent, with considerable differences in prevalence from one country or region to another.

The rate of extreme poverty is thus close to zero in most of the countries of North Africa, and is relatively low in several countries of Southern Africa (except for Lesotho and Malawi). On the other hand, the rest of sub-Saharan Africa is poor on the whole, despite large differences here, too: the regions of Central Africa, the Gulf of Guinea and the Indian Ocean have the highest poverty rates (see map).

The demographic equations are such that the number of people living in situations of extreme poverty is highly concentrated in a few countries with large populations. Over half of the 469 million poor people at present come from just five countries, even though they do not have the highest poverty rates on the continent: Nigeria, the Democratic Republic of the Congo, Ethiopia, the United Republic of Tanzania and Madagascar.

A MULTIFACETED CONTINENT WITH SHARED CHALLENGES

Fragility and development: two interconnected challenges

A CONTINENT STILL MARKED BY FRAGILE SITUATIONS

Over half of the countries that are in a situation of fragility, meaning those whose public authorities are not able to provide the basic services and security necessary for the population throughout their territory, are in Africa. According to the World Bank, the continent counts 20 fragile countries out of the 36 around the world (see map).

This classification into two categories no doubt conceals many more complex and diversified realities. It does serve, however, to identify those states that are having the greatest difficulty pursuing a common national interest, and which therefore have the highest risks of political and economic instability.

The fragile states in Africa are home to 200 million people, representing some 20% of the population of the continent.

These are the same countries that have the highest poverty rates. Consequently, the maps showing the levels of poverty and violence can be aligned with that for fragility (see pp. 46–47).

AN AXIS OF FRAGILITY ACROSS THE CONTINENT

These situations of fragility do not affect the whole of the continent. A very clear vertical axis runs from Libya down to Zimbabwe, via Chad, the Central African Republic and the Democratic Republic of the Congo.

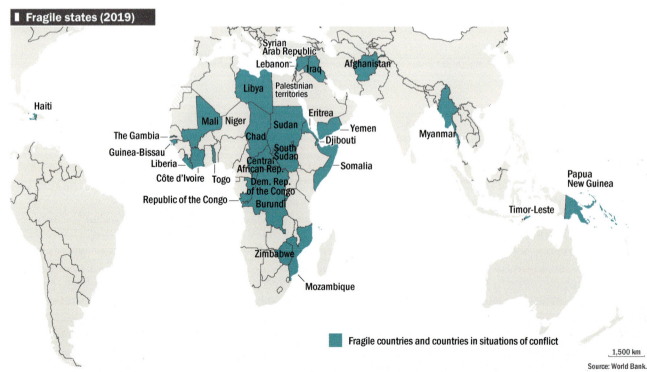

Fragile states (2019)

Fragile countries and countries in situations of conflict

Source: World Bank.

The World Bank ranking is drawn up on the basis of a Country Policy and Institutional Assessment (CPIA) indicator. The countries considered to be in a situation of fragility or conflict are those whose CPIA indicator is less than or equal to 3.2, which includes countries emerging from crises or conflicts or countries that have seen the presence of (regional or United Nations) peace-keeping or mediation operations on their territory within the past three years.

A horizontal axis is also appearing owing to the deteriorating situation in Burkina Faso and increasing violence against civilians in Somalia and Sierra Leone. This regional concentration provides a clue to the multiple factors in this fragility, which relate to conflicts, governance problems, weak institutions, issues of access to natural resources and poor economic performance, in particular when the territory of the country is very extensive.

REFUGEES: THE INTERNATIONAL FACES OF FRAGILITY

Whether they take the form of violence, famine or exclusion, these situations of fragility result in the population of these zones fleeing. According to the UN Agency for Refugees (UNHCR), there were 17.8 million displaced people in Africa in 2018, 90% of whom were concentrated in six countries, the Democratic Republic of the Congo, Somalia, Ethiopia, Nigeria, South Sudan and Sudan.

The continent was also home to 7.4 million refugees, i.e. people who have fled their country, two-thirds of whom came from four areas of armed conflict, which is to say South Sudan (2.3 million), Somalia (950,000), the Democratic Republic of the Congo (720,000) and the Central African Republic (590,000). Almost 90% of these exiles have found refuge in another African country, and a neighbouring country in almost all cases. Uganda and Sudan were thus the third- and fourth-ranking countries in the world in 2018 for the number of refugees hosted there, after Turkey and Pakistan.

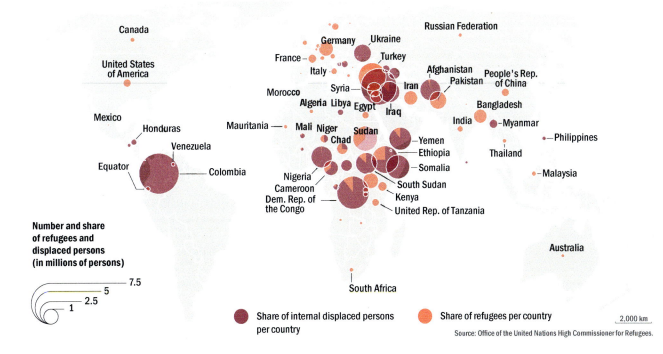

Refugees and displaced persons around the world (2018)

Source: Office of the United Nations High Commissioner for Refugees.

Inequalities: a situation of contrasts

FROM CAIRO TO CAPE TOWN, INCREASING INCOME INEQUALITY

Statistically speaking, the continent of Africa is second in the world in terms of inequalities, after South America. At least that is what would appear to be suggested by the Gini index, a statistical measurement showing the extent of wealth inequality within a country's population, when applied to income in Africa.

Inequality indicators are always awkward to handle on a continent where little is known about the distribution of income and expenditure and where information on assets is almost non-existent. The data that are available would appear to indicate that there are great disparities in the extent of inequality, with prevalence clearly increasing as we move from north to south.

North Africa and the countries of the Sahel thus appear to be the regions where the gaps in revenues between individuals are the least pronounced, while Central Africa and, above all, Southern Africa, contain the majority of the most inegalitarian countries.

NORTH AND SOUTHERN AFRICA: WHY ARE THERE SUCH DIFFERENCES IN INEQUALITY?

It is essential to take account of inequalities because, in Africa as elsewhere, the average level of wealth of a country provides no indication as to the way in which that wealth is distributed. The comparison between the countries of North Africa and those of Southern Africa is most revealing in this respect.

These two regions, the richest in Africa, have similar average levels of GDP per capita (around $10,000 annually), but the extent of inequality is very different there. Algeria and Egypt are thus among the countries with the lowest measured levels of inequalities on the continent, while they are at their highest in South Africa, Botswana, Namibia and Zambia.

In North Africa, the state has long fulfilled a redistribution function via a system of subsidies which has made it possible to reduce the differences in income to some extent. In Southern Africa, the level of inequalities has been accentuated, on the contrary, by the legacy of apartheid and by a mining resource exploitation-driven growth model with little focus on redistribution.

THE IMPACT OF INEQUALITY ON THE EROSION OF SOCIAL TIES IN AFRICA

High monetary inequality is likely to undermine the social ties that are the very foundation of stability in our societies. In return, low social cohesion makes it more difficult to reach political consensus in favour of implementing redistribution policies. In addition, the perception individuals have of their own situation counts more than the reality that is actually measured. Measurements of happiness and well-being thus show a different face of Africa, with the poorest countries not necessarily being the unhappiest (see map).

This dimension is one that must be taken into account: in North Africa, for example, the growing and generalized feeling among the population of dissatisfaction with their quality of life is now considered to have been one of the causes of the Arab Spring in 2011. This difference between objective data and opinion data in North Africa is partly due to inequalities in opportunities, in particular in terms of access to employment and to good-quality public services.

These inequalities are multidimensional and complex: they may be social (access to essential services, such as education, health and housing), political (inequalities in participation in political life, for example), territorial (gaps between town and country), or related to gender (see pp. 88–89).

Overview of inequalities (2017)

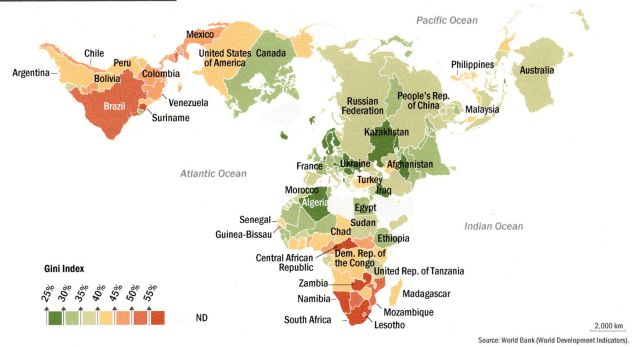

Source: World Bank (World Development Indicators).

The Gini index is a statistical measurement showing the distribution of revenue and expenditure within a population. The coefficient varies from 0 (perfect equality) to 100 (extreme inequality).

Perceived happiness in Africa (2018)

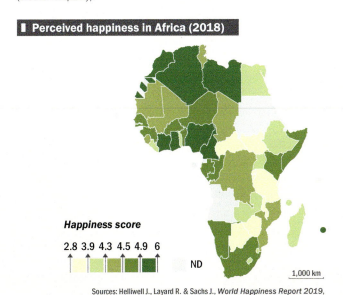

Sources: Helliwell J., Layard R. & Sachs J., *World Happiness Report 2019*, New York, Sustainable Development Solutions Network, 2019.

The happiness index is based on data from the Gallup World Poll and more precisely those responses relating to perceptions of quality of life. The higher the indicator, the greater the perception of happiness.

A view of the township of Philippi in Cape Town, South Africa. Philippi is one of the largest slums in the region and is home to thousands of families living in poor conditions. Philippi lies alongside the Hawequas mountain chain and is built around busy roads into the centre of Cape Town. Photo © Andrew Aitchison/Alamy. Stock Photo.

Urbanization: a strong trend of settlement in Africa

URBANIZATION ACCELERATING IN AFRICA

Africa's cities are booming. With an urbanization rate of 43% in 2018, compared to 14% in 1950, the continent is in the throes of a strong phase of urbanization, similar to that witnessed in Asia in recent decades.

The growth rate of Africa's urban population is one of the highest in the world, at 4% per year on average between 1950 and 2018. By 2050, the urban population of Africa will thus be the second largest in the world. By then, almost 1.5 billion people will be living in cities, a lower number than in Asia (3.5 billion), but much higher that the number of city-dwellers in Europe (600 million), Latin America (685 million) or North America (387 million).

THE MULTIPLICATION OF MEGACITIES

This phenomenon of urbanization is driving the appearance of large urban areas: the number of metropolises in Africa was thus grew by a factor of 10 between 1970 and 2015.

The largest cities, and in particular the national and regional capitals, are undergoing spectacular development. Between 1985 and 2025, the population of the 10 largest African cities should grow from 28.7 million to over 107 million. By 2050, two of the world's most populous cities (Kinshasa and Lagos) will be in Africa, and four by 2075 (Kinshasa, Lagos, Dar es Salaam and Cairo). In 2100, if such growth rates continue, the world's three biggest cities will be African. By that date, Lagos in Nigeria, Kinshasa in the Democratic Republic of the Congo and Dar es Salaam in the United Republic of Tanzania will have become megacities that are home to 88, 84 and 74 million people, respectively.

RURAL AREAS AND SECOND-TIER CITIES: GENUINE CENTRES OF URBANIZATION IN AFRICA

Despite the future size of the African megacities, it is elsewhere that much of this urbanization will occur. These cities with over one million inhabitants will represent only 45% of the growth in the African population through to 2025, with the other 55% being in intermediate-sized urban areas of less than one million people (see map).

African urbanization will therefore take place above all through the extension of middle-sized cities into the rural areas around them, thereby reinforcing the "rurbanization" effect across the continent. The secondary cities, which extend their urban areas into rural zones, will thus play the role of a mid-way space between the megacities and rural areas. This phenomenon is already at work today and will soon become of structural importance throughout the continent. Currently, 81% of Africans live in a rural zone-urban zone interface and they will be the first to be concerned about the development of second-tier cities.

The connection between agricultural production zones and urban consumption zones could thus constitute a driver for the elimination of rural poverty and an improvement in the quality of life in urban areas. Smoother mobility between cities and countryside could ultimately be conducive to greater economic dynamism and better access for those living in rural areas to centralized services (administration, hospitals, banks, etc.).

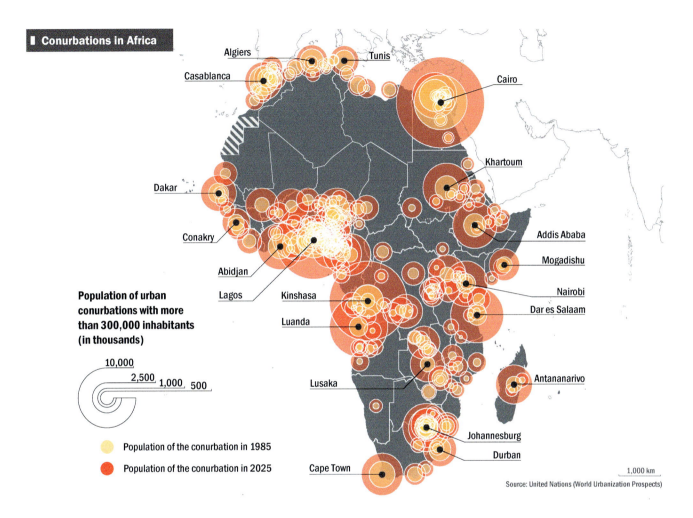

Conurbations in Africa

Population of urban conurbations with more than 300,000 inhabitants (in thousands)

- Population of the conurbation in 1985
- Population of the conurbation in 2025

Source: United Nations (World Urbanization Prospects)

The 10 largest cities in the world

#	2010 City	Population (in millions)	2025 City	Population (in millions)	2050 City	Population (in millions)	2075 City	Population (in millions)	2100 City	Population (in millions)
1	Tokyo	36	Tokyo	36.4	Bombay	42.4	Kinshasa	58.4	Lagos	88.3
2	Mexico City	20.1	Bombay	26.3	Delhi	36.1	Bombay	57.8	Kinshasa	83.5
3	Bombay	20	Delhi	22.49	Dacca	35.2	Lagos	57.2	Dar es Salaam	73.678
4	Beijing	19.6	Dacca	22	Kinshasa	35	Delhi	49.3	Bombay	67.24
5	São Paulo	19.5	São Paulo	21.4	Calcutta	33	Dacca	46.2	Delhi	57.334
6	New York	19.4	Mexico City	21	Lagos	32.6	Calcutta	45	Khartoum	56.594
7	Delhi	17	New York	20.6	Tokyo	32.6	Karachi	43	Niamey	56.149
8	Shanghai	15.7	Calcutta	20.5	Karachi	31.7	Dar es Salaam	37.4	Dacca	54.25
9	Calcutta	15.5	Shanghai	19.4	New York	24	Cairo	32.9	Calcutta	52.395
10	Dacca	14.7	Karachi	19	Mexico City	24.3	Manilla	32.7	Kabul	50.27

Source: Hoornweg D and Pope K., 2016, "Population Predictions for the World's Largest Cities in the 21st Century", *Environment & Urbanization*, International Institute for Environment and Development (IED), 29(1): 195–216.

Simultaneous densification of cities and countryside

A FEATURE OF AFRICA: SIMULTANEOUS SETTLEMENT OF CITIES AND COUNTRYSIDE

The settlement of Africa has followed a different path from that observed elsewhere in the world: it is characterized by rapid growth not only in the urban but also in the rural population, revealing the absence of any large-scale rural exodus in Africa.

The rate of growth of the rural population is no doubt a unique feature of Africa. The rural population grew by a factor of four in Africa between 1950 and 2015, while it only doubled in Asia during the same period. The rural population should therefore continue to grow at a rate in excess of 1% a year on average through to 2050, while the number of people living in rural areas in Asia peaked in 2002 and has been declining since then.

Africa will also be the only continent where the rural population is continuing to grow, while still representing less than half of the total population of the continent in 2023.

DIVERSE FORMS OF SETTLEMENT IN AFRICA FROM ONE REGION TO ANOTHER

Within this overall trend, significant disparities can be observed, since the regions of Africa are not all experiencing the same growth dynamics in their urban and rural populations.

While North Africa and Southern Africa have commenced their urban transition, with the share of the rural population now less than the number of city-dwellers, West Africa and East Africa are likely to see both their urban and rural populations growing at the same time.

In addition, regional differences aside, the landlocked countries on the continent are going to remain mainly rural through to 2050 and beyond. Whether they are in Central, East or Southern Africa, the urban population of these countries will still represent less than half of their total population.

DENSIFICATION OF INHABITABLE SPACES

Simultaneous growth of the urban and rural populations will contribute to overall densification of all the inhabitable spaces on the continent (see map). Between 1950 and 2016, the density of the African population increased fivefold, from eight to 40 inhabitants per sq km. Africa has thus caught up with and overtaken North America and Latin America, and this densification is set to continue.

Unlike in Europe, the population of Africa is spread unequally over the continent, even when allowances have been made for its desert areas (Sahara and Namibia). The medium-term consequences of these demographic trends are therefore very specific to each region. The coastal areas of North, West and East Africa, which are already the most densely populated, are those where the management of this densification of the territory will be the most immediate. In those regions that are vulnerable to climate change and where tensions around natural resources are already severe (in the Sahel, for instance), anthropic pressure will be more acute. As for Central Africa, which is more sparsely populated, the densification of habitable spaces may progress to the detriment of forest cover.

Highway leading out of Ouagadougou, Burkina Faso.
Photo © MattLphotography/Shutterstock.

Population density around the world

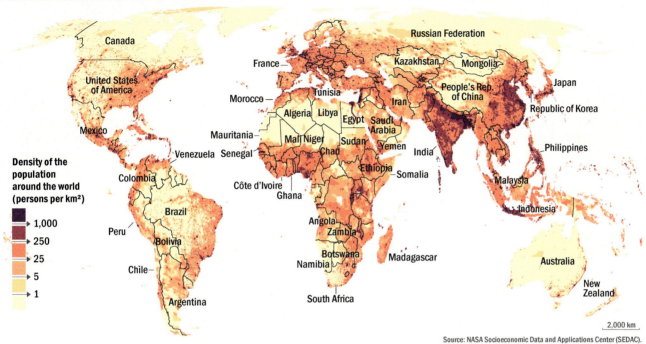

Source: NASA Socioeconomic Data and Applications Center (SEDAC).

Trends in urban and rural populations in Africa (left) and Asia (right)

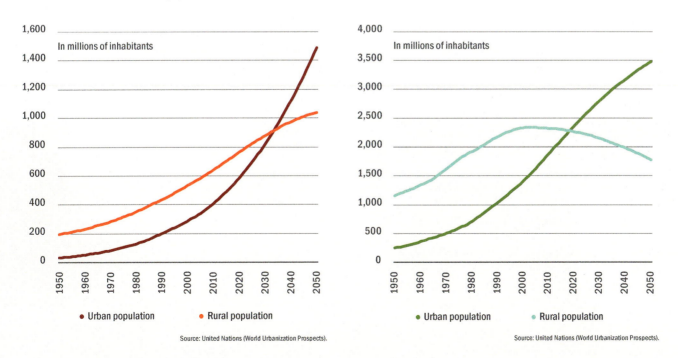

Source: United Nations (World Urbanization Prospects).

Rural and landlocked zones: the main factors in unequal access to basic services

HUGE DIFFERENCES IN ACCESS BETWEEN URBAN AND RURAL ZONES

The urban and rural structure of Africa is one of the most obvious keys to understanding development disparities on the continent. The gap between urban and rural spaces can be seen first of all in poverty rates, which are twice as high on average in rural areas than in the cities. It can also be seen in the considerable difference between rates of access to basic services, testifying to territorial inequalities between cities and rural areas, sometimes on a scale not found anywhere else in the world.

For instance, while 81% of African city-dwellers had electricity in 2017, only 35% of those living in rural areas had such access (see graph). This difference in access in fact concerns all those services that are conducive to an improvement in the living conditions of the population, such as health, sanitation and education, the distribution of which is largely unfavorable to the rural areas.

A DIFFICULT HERITAGE TO SHED, BUT SOME ENCOURAGING EXAMPLES

The origin of this development gap between urban and rural areas no doubt partly lies in the challenges that drove the formation of the post-colonial states. After independence, the central spaces—and notably capital cities—were often better treated with the aim of establishing the new powers there and building the administrative and economic centres of their young nations. The effort to build infrastructures was thus accentuated in the largest urban centres.

In a context of limited financial resources and concentration of economic activities in the large urban centres, it can prove difficult to overcome this legacy of territorial polarization to the detriment of the countryside. The construction of rural infrastructures, administrative decentralization, financing of local authorities and territorial innovation are thus key to reducing inequalities.

The situations do vary on the continent, however, and certain regions have relatively better territorial coverage than others. Although these disparities do continue, sometimes in a pronounced manner, urban/rural disparities are smaller in the richest countries on the continent, such as South Africa, Egypt or Tunisia.

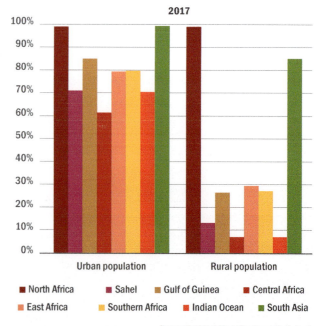

■ **Unequal access to electricity**

Source: World Bank (World Development Indicators).

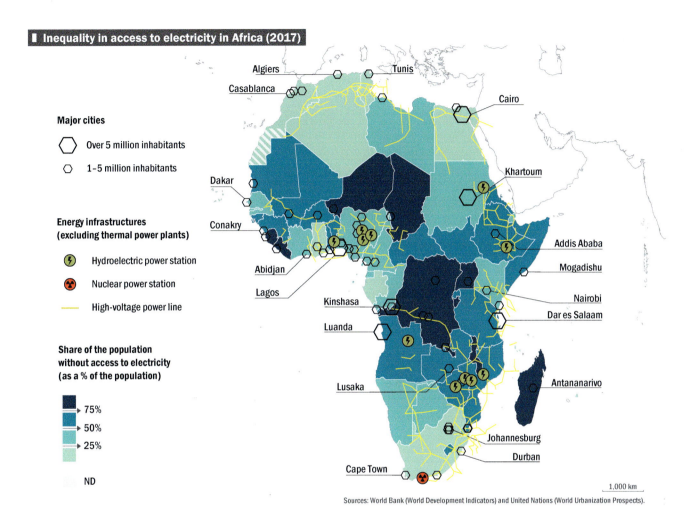

A LANDLOCKED GEOGRAPHICAL POSITION REMAINS AN OBSTACLE TO ACCESS

Another major explanatory factor for the differences in access to services and infrastructures is whether a country or territory is landlocked. Those that do not have coastal areas appear on average to be less well equipped in infrastructures, whether roads, electricity or digital access (see map). Genuinely transcontinental infrastructures are in fact limited, which restricts intra-African exchanges.

One of the explanations for this situation is linked to the economic history of Africa: part of the infrastructures were built to facilitate exports of commodities to the rest of the world, from the interior of the continent to the major ports. Other factors also come into play, such as the greater fragility of the countries in the centre of Africa, the concentration of the population in small parts of the territory, or issues of cost, financing and economic balance.

Solutions are already being rolled out, however. In electricity, for example, the development of high-voltage lines—facilitating interconnections between grids—or standalone grids are powerful drivers of regional integration and development of access to electricity in landlocked countries or regions.

Economic integration progressing in Africa

REGIONAL ORGANIZATION OF THE CONTINENT

Like the other continents, Africa has engaged in a process of regional economic integration. Founded in 2002 to take over from the Organization of African Unity (OAU), one of the purposes of the African Union (AU) is to promote regional integration, and to that effect it recognizes several Regional Economic Communities (RECs).

There are eight of these intergovernmental organizations, created between groups of neighbouring countries with the aim of reinforcing their economic cooperation: the Inter-Governmental Authority for Development (IGAD), the East African Community (EAC), the Southern African Development Community (SADC), the Community of Sahel–Saharan States (CEN–SAD), the Economic Community of Central African States (ECCAS), the Economic Community of West African States (ECOWAS), the Common Market for Eastern and Southern Africa (COMESA) and the Arab Maghreb Union (AMU) (see map). These entities ultimately cover all the African countries. Some countries are members of more than one of them, thereby creating overlaps between the RECs.

INTEGRATION PROCESSES ENGAGED TO VARYING DEGREES

The creation of these communities has facilitated undeniable progress in terms of regional integration, even though the degree to which they have advanced varies from one region to another.

Five of the eight RECs have thus established free trade areas, the first step towards economic integration. Three of them—EAC, ECOWAS and COMESA—have pushed their degree of integration further by creating a customs union. The free movement of people has also improved in the majority of the RECs (see pp. 66–67) and all the members of ECOWAS apply this protocol.

Economic integration is also very much a reality in the other African sub-groups, with monetary integration in the cases of the West African Economic and Monetary Union (WAEMU), the Central African Economic and Monetary Community (CAEMC) and the Southern African Customs Union (SACU). The countries that belong to these organizations share the same currencies, coordinate their economic policies and are extensively engaged in advanced regional economic cooperation.

Integration status by regional economic community

REC	Free trade area	Customs union	Single market	Country that has applied the protocol on the freedom of movement of persons	Economic and monetary union
EAC	Yes	Yes	Yes	3 countries out of 5	No
COMESA	Yes	No	No	Only Burundi has ratified the protocol; Rwanda is expected to do so.	No
ECOWAS	Yes	Yes	No	All 15 countries	No
SADC	Yes	No	No	7 countries out of 15	No
ECCAS	Yes	No	No	4 countries out of 11	Yes.
CEN-SAD	No	No	No	Situation not clear	No
IGAD	No	No	No	No protocol	No
AMU	No	No	No	3 countries out of 5	No

Source: Economic Commission for Africa (2016).

AN INTEGRATION MEGA-PROJECT: THE AFRICAN CONTINENTAL FREE TRADE AREA (AfCFTA)

The most recent step forward in economic integration in Africa occurred not at the regional level, but at the level of the whole continent. The African Union has obtained agreement from all its members for the implementation of the AfCFTA. The ambition of this trade agreement, which brings together all the African states, is to set up the world's largest free trade area since the creation of the World Trade Organization (WTO) in 1995. Signed by a majority of countries in March 2018, it entered into force on 30 May 2019 after the submission of the ratification instruments of 22 countries (minimum threshold set by the agreement).

Implementation of the free trade area is moving forward progressively as customs duties are gradually eliminated on intra-African trade. If the project is taken through to full completion, the AfCFTA will constitute a market of 1.2 billion people, representing GDP of $2,500 billion.

Development of intra-continental trade is a major challenge for the economic future of the continent (see pp. 60–61). The Economic Commission for Africa (ECA) thus estimates that the AfCFTA could boost intra-African trade by 52.3% by eliminating customs duties on imports. Another challenge will also be to reduce non-tariff barriers (administrative procedures and formalities, standards, poor infrastructures).

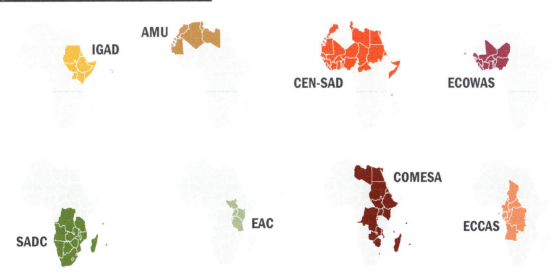

Regional Economic Communities (RECs)

Sources: The Economic Commission for Africa, the African Union and the African Develpment Bank (2017), *Assessing Regional Integration in Africa VIII*.

Growth in intra-African trade as a driver of diversification

EXPANDING TRADE WITHIN THE CONTINENT

Over the past two decades, intra-African flows of goods have increased rapidly, as Africa has increasingly been integrated into the world trade system. The share of intra-African trade in total trade on the continent is now close to 14.5%, compared to 11.6% in 2008. This share is ultimately quite similar to that observed in other emerging and developing regions of the world, such as the Pan-Arab Free Trade Area (PAFTA) or the Latin American Integration Association (LAIA), for example.

This commercial integration process hides sharp contrasts from one country or sub-region to another, however. The expansion of regional trade has driven the emergence of trading hubs, notably South Africa in the southern part of the continent, Côte d'Ivoire and Senegal in West Africa, and Kenya in East Africa. On the other hand, some of the large African countries are still poorly integrated into the continent's trade more generally. For instance, trade between the countries of North Africa and the rest of the continent only represents 13.4% of total intra-African trade, despite this region representing more than one-fourth of Africa's GDP (26.9%) all on its own.

TRADE UNDERESTIMATED ON ACCOUNT OF THE VALUE OF INFORMAL CROSS-BORDER TRADE

The growth in intra-African trade only partly reflects the integration that is underway. Its scale is underestimated because the statistics do not always take account of informal trade flows (foodstuffs, hydrocarbons, vehicles, etc.) that cross the continent's borders.

The share of informal activities in foreign trade varies from one country to another, depending on the products and other specific circumstances. Their estimated volume is considerable, however.

According to the International Food Policy Research Institute (IFPRI), informal exports are likely

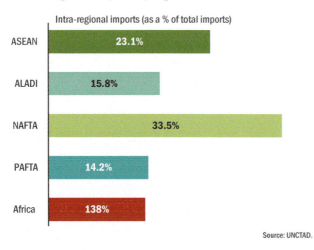

Intra-regional imports by region

Source: UNCTAD.

ASEAN: Association of Southeast Asian Nations
LAIA: Latin American Integration Association
NAFTA: North American Free Trade Agreement
PAFTA : Pan-Arab Free Trade Area

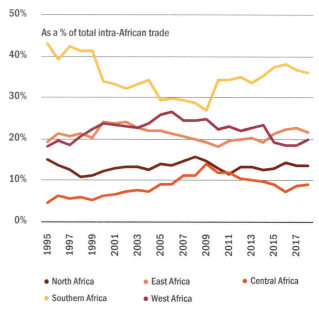

Intra-African trade

Source: UNCTAD.

to have represented between 14% and 34% of Uganda's exports between 2008 and 2018, for example (see graph). Within the South African Development Community (SADC), informal trade in certain foodstuffs represented 30% to 40% of official trade in the 2000s. In Benin, informal exports of goods (especially of agricultural products) to Nigeria could be five times greater than the volumes recorded by customs.

INTRA-AFRICAN TRADE: A FACTOR ACCELERATING GROWTH

Growth in intra-African trade is driving considerable gains in development, as these flows are more diversified than those with the rest of the world. More particularly, intra-African exports incorporate twice the amount of manufactured goods (over 40%) than do exports to the rest of the world (20%).

Intra-African trade thus contributes to the up-market shift of the continent's production sector. It constitutes another way of reinforcing its capacity to benefit from its own market. A recent study by the International Monetary Fund (IMF) estimates that African countries gain an average of 0.5% growth in GDP each time intra-regional exports increase by 5%.

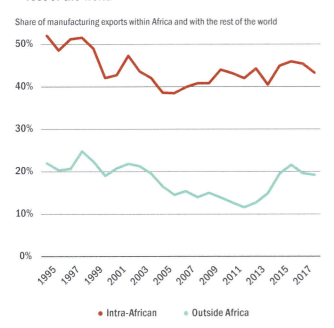

Manufacturing exports within Africa and with the rest of the world

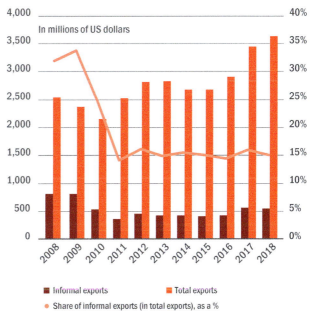

Total and informal Ugandan exports

Source: Uganda Bureau of Statistics, 2019.

Emergence of a pan-African financial system

INTRA-CONTINENTAL INVESTMENTS ON THE RISE
Who are the main foreign investors in the countries of Africa? Increasingly, they are African investors, who are looking for investment opportunities in countries on the continent other than their own. According to the African Development Bank (AfDB), intra-continental foreign direct investment (FDI) in greenfield projects (creation of new production units) more than doubled between 2006 and 2016—from $4 to $10 billion—while the number of intra-continental merger and acquisitions transactions increased from 238 to 418 during the same period. These investments are driven mainly by African multinationals, such as Nigerian construction group Dangote and Ethiopian airline company Ethiopian Airlines, within the framework of their regional or pan-African expansion strategies.

Ultimately, three countries (Morocco, South Africa and Egypt) were among the 12 main investors in Africa in 2015–2016. In the same year, Morocco, the intra-continental investment champion, invested more heavily in Africa than countries such as France or the United Kingdom (see graph).

DEVELOPMENT OF A FINANCIAL INDUSTRY
The development of intra-African financial flows is also driven by a network of pan-African financial institutions that is constantly growing. In recent years, several African banking groups have thus adopted a continental expansion strategy to establish themselves in their regional spaces and beyond. Banks from Morocco (Attijariwafa Bank), Kenya (Kenya Commercial Bank–KCB Group, in particular), Nigeria (United Bank of Africa), South Africa (Standard Bank) and Togo (Ecobank) have all opened subsidiaries in other African countries.

The multinational financial institutions are contributing to the financial integration of the continent and are establishing connections across borders between African customers in search of financing and those capable of providing it. Their development makes it possible for African groups to share their expertise with the rest of the continent and to reinforce the global banking industry. Ultimately, the financing of the real economy in the countries where they are established is enhanced (see pp. 108–109).

THE EMERGENCE OF REGIONAL FINANCIAL CENTRES
The financial markets are also playing an increasingly important role in the creation of a continental financial system. States, and also the private sector, benefit from these inputs of capital, and more particularly in the financial services and telecommunications sectors.

African stock markets have proliferated throughout the continent since the 1990s, and a certain number of them have a regional presence. This is the case notably of the market for regional securities in Abidjan, which enables investors to trade the securities of companies throughout the WAEMU area.

Another core trend is the proliferation of private equity funds specializing in the management of portfolios of African companies in their regions, from their bases in Côte d'Ivoire, Kenya, South Africa and Morocco. These investment funds are mostly based in Africa and are medium-sized entities that have in-depth knowledge of their local contexts, conduct explorations of their markets and monitor performances.

This channel has developed considerably in recent years and has raised an average of almost $3 billion per year from investors since 2010.

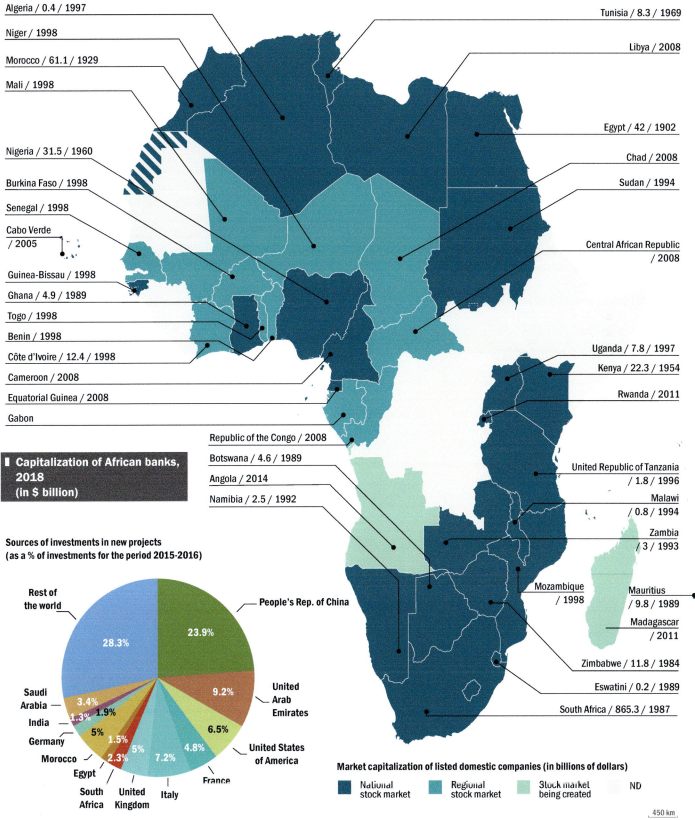

Intra-continental migration: human ties between African areas

MAINLY INTRA-CONTINENTAL MOBILITY

The share of Africans living in a country other than their country of origin represented 14.7% of the world's migrant population in 2019, which is a much smaller proportion than that observed for the populations of Asia and Europe (41% and 23%, respectively). Africa is also the main destination for African immigrants. In 2019, more than half of the 40 million international migrants coming from an African country (21 million people) emigrated within the continent. Their number was therefore almost twice as high as in Europe, which received 11 million migrants from Africa that year (see map).

The main regions from which over half of the migrants come are North Africa and the Horn of Africa, followed by the Gulf of Guinea and Southern Africa. The main regions to which they go are East Africa and West Africa. Five destination countries receive almost one-fourth of the African migrants in Africa: South Africa (2.3 million), Côte d'Ivoire (2.5 million), Uganda (1.7 million), Ethiopia (1.2 million) and Nigeria (1.1 million).

WHY DO AFRICANS MIGRATE?

The data available indicates that insecurity is not the primary cause of migration within Africa. In fact, only 20% of the migrants are refugees. According to surveys of the population carried out by the Afrobarometer network, it is the hope of better economic and social prospects that is the main driver of these population movements, with the main reasons that are mentioned being, by order of importance, to find a job or better employment opportunities, and to flee difficult economic conditions and poverty.

Most of these African migrants are young people under the age of 30 who have a relatively good level of education and are in search of a job. Almost half of them are women.

PROXIMITY AS A DETERMINANT OF MIGRATION

While migrants coming from the countries of North Africa emigrate mainly outside the continent—over half of them towards Europe (57%)—almost 70% of sub-Saharan African migrants remain in Africa. They migrate essentially towards the countries that are seen as the regional economic hubs, such as South Africa in the southern part of the continent and Côte d'Ivoire in West Africa, and which are also the closest geographically. Ultimately, over three-fourths of international migrants coming from these regions migrate within the region.

Geographical proximity therefore plays a decisive role, which can be explained notably by the introduction of free movement systems in the Regional Economic Communities (RECs) and by the porous nature of certain borders (as illustrated by the migration movements within the Gulf of Guinea and the Sahel), which facilitate movements from one country to another.

African migrant workers transported on the back of a dump truck, South Africa. Photo © Chadolfski/Shutterstock.

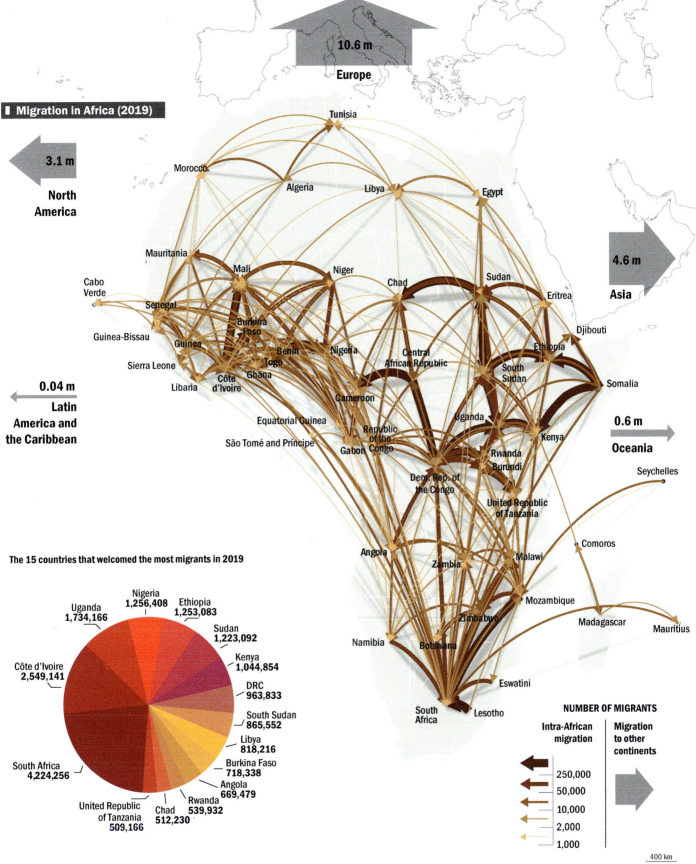

Human mobility taking shape on the continent

BARRIERS TO MOBILITY IN AFRICA

International mobility is limited for the citizens of Africa. Of the 50 passports around the world that offer the fewest possibilities in terms of entering foreign countries without requiring a visa, 28 are issued by countries in Africa. These restrictions also concern movements within the continent.

According to the Visa Openness Index of the African Union Commission (AUC) and the African Development Bank (AfDB), generally African nationals can enjoy visa-free travel in one-fourth of the other countries on the continent and they can obtain a visa on arrival in 26% of the other African countries. They are therefore required to apply for a visa prior to their departure to half of the countries (see graph).

Mention should be made, however, of some progress in visa openness in Africa since 2016. In 2019, for the first time African travellers had access, with or without a visa on arrival, to over half of the countries on the continent. The admission rules vary according to the country of origin, with some nationalities being more welcome than others, but as a general rule there is less freedom of movement in Africa than anywhere else.

PROMOTING REGIONAL FREE MOVEMENT

The absence of free movement is considered detrimental to the continent's development, as it hinders the mobility of skills on the labour market, the development of tourism and dissemination of knowledge. It is for this reason that the African Union and several countries are keen to foster it, primarily via the regional organizations.

Certain Regional Economic Communities (RECs) have made particularly strong progress in this area, by establishing the right of Africans to travel within their group of countries and to settle elsewhere in Africa.

More particularly, the East African Community (EAC) and the Economic Community of the West African States (ECOWAS), each of which include 11 of the 20 most open countries on the continent, made a major breakthrough by adopting a regional free movement treaty and creating a common passport, in 2009 and 2014, respectively. The free movement area of the Southern African Development Community (SADC) is not yet fully operational, however, as it has not been ratified by all the states, but negotiations on the subject have facilitated the multiplication of bilateral visa-exemption agreements.

TOWARDS A SINGLE PASSPORT FOR ALL AFRICANS?

Africa dreams of establishing freedom of movement across the continent. The project of free movement throughout the territory is a legacy of long-standing pan-African aspirations and is ready to be implemented in technical terms.

The African common passport was launched in July 2016 and the African Union protocol on the free movement of persons, the right of residence and the right of establishment in Africa was adopted in January 2018. However, the ratification process is still encountering some reluctance among the states, notably on the issue of controlling migration, refugee flows and health risks, which by nature know no borders, as was shown, for example, by the Ebola epidemic in West Africa in 2014.

In the meantime, several countries have decided to opt for total pan-African mobility, such as Rwanda, Seychelles and more recently Benin, which have made it possible for all African nationals to enter their territories without a visa for periods of less than 90 days.

▌ Most open African countries in terms of visas in 2019

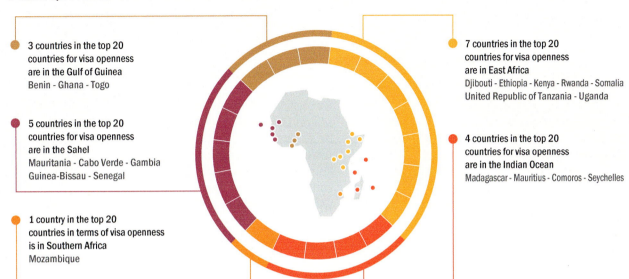

- 3 countries in the top 20 countries for visa openness are in the Gulf of Guinea
  Benin - Ghana - Togo

- 5 countries in the top 20 countries for visa openness are in the Sahel
  Mauritania - Cabo Verde - Gambia Guinea-Bissau - Senegal

- 1 country in the top 20 countries in terms of visa openness is in Southern Africa
  Mozambique

- 7 countries in the top 20 countries for visa openness are in East Africa
  Djibouti - Ethiopia - Kenya - Rwanda - Somalia United Republic of Tanzania - Uganda

- 4 countries in the top 20 countries for visa openness are in the Indian Ocean
  Madagascar - Mauritius - Comoros - Seychelles

Source: Visa Openess Index.

▌ Openness in terms of visas in 2019

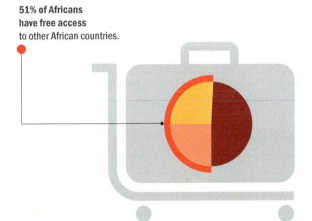

51% of Africans have free access to other African countries.

- 25% of Africans do not need a visa to travel to other African countries.
- 26% of Africans can obtain a visa on arrival in other African countries.
- 49% of Africans need a visa to travel in other African countries.

Source: Visa Openess Index.

A young man embracing his girlfriend at the airport.
© PeopleImages/iStock by Getty Images.

Creative industries: the face of an emerging cultural industry

THE RISE OF AFRICAN CINEMA

Although its international presence is limited, movie production is thriving on the continent. Over the past 20 years, the number of film-producing countries and the number of movies produced have increased sharply (see map). This rise has been largely driven by Nigeria, whose "Nollywood" movie industry (a reference to Hollywood in the United States and Bollywood in India) has become one of the most flourishing in the world.

With some 2,000 movies produced each year, Nollywood has become the second largest cinema industry in the world, after that of India. Nigerian cinema is booming and now represents 2% of national GDP and 300,000 direct jobs. Several other countries are also big movie producers, along with more traditional countries such as Egypt and the countries of the Maghreb, and the emergence of new players, such as Ghana, Mozambique and Uganda.

The continent is also home to several renowned movie festivals, such as the Pan-African Film and Televisions Festival (FESPACO) which is held every two years in Ouagadougou, Burkina Faso.

CULTURAL AND CREATIVE INDUSTRIES: A WINNER FOR AFRICA

Movies are not the only creative sector that is booming on the continent. Progress can be seen generally in all the cultural and creative industries (CCIs): in music, radio, design, photography, crafts, media, publishing and also video games. South Africa is the country where the music industry is the most mature, with an average of 550 albums produced each year and several plaudits at the MTV Africa Music Awards.

The cultural diversity of an increasingly connected African population and its youth are two factors that are likely to contribute to the development of a sector that is often underestimated. Demand for cultural goods that are produced locally can make the CCIs a source of income and jobs, while certain activities, such as fashion, music festivals and architecture, can foster tourism and contribute to making urban areas more attractive.

NATIONAL CULTURAL CONTENT SHARED WIDELY THROUGHOUT AFRICA

Cultural influences at a regional level are becoming stronger. The internet and mobile phones have considerably modified modes of production and dissemination. Making up for the lack of cinema screens, these new channels can provide extensive visibility for African cinema.

To finance production costs in a price-sensitive market, African television, film and music producers seek to maximize audiences by selling beyond their domestic borders. Nigerian movies and soap operas, either in their original or adapted versions, are therefore much appreciated in Central and Eastern Africa. As proof of their popularity also in French-speaking West Africa, the Nollywood television channel has the region's biggest audience. The video game market is also developing.

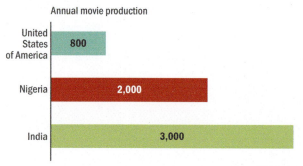

▌ Annual movie production in India, Nigeria and the United States

Annual movie production

- United States of America: 800
- Nigeria: 2,000
- India: 3,000

Source: UNESCO (2015).

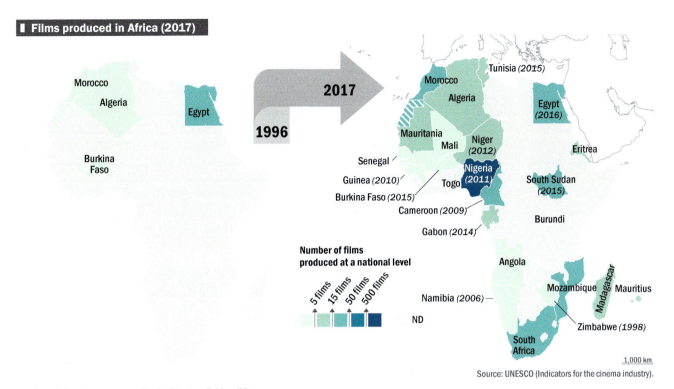

Films produced in Africa (2017)

The dates in brackets correspond to the latest available edition.

Source: UNESCO (Indicators for the cinema industry).

In Kenya and Ghana, the company Leti Arts produces cartoon strips and video games for mobile phones, such as *Ananse: The Origin*, based on a folk epic from Ghana. It has also developed *Afrocomix*, a mobile app that gives access to different content (comic strips, wallpapers, short animations, etc.) inspired by the tales of African culture (Cameroon, Ghana, Kenya, Egypt, Senegal, South Africa, Zimbabwe, etc.) and destined for the whole of the continent and its diaspora.

Nollywood. Ini Edo (left) and Halima Abubakar film a scene in Lagos, Nigeria, on 26 August 2009.

Photo © Shashank Bengali/MCT/Tribune News Service via Getty Images.

Africa, a blue continent

LITTLE-KNOWN RICHES IN WATER

Africa is a continent rich in water and groundwater resources. The river network there comprises almost 17,000 water courses of at least 100 km^2, representing about 20% of those on the planet as a whole—a proportion that is equivalent to Africa's share of the world's land surface area. The surface area of Africa's lakes is in excess of 200,000 km^2 in total, representing about 13% of the world's lake surface area.

It is also estimated that the renewable groundwater resources of Africa constitute 9% of the world's fresh water resources, while there are also large reserves of potable water (660,000 billion m^3) that are not renewable, however, on the continent as a whole. Over two-thirds of the countries of Africa (38 out of 54) are coastal countries that share a coastline of 30,750 km and 13 million km^2 of maritime areas under African jurisdiction.

WATER: A COMMON GOOD TO BE SHARED BETWEEN NATIONS

These networks and spaces are unevenly distributed across the continent and draw borders that are different from the political and administrative divisions, making sustainable management of water resources more complex. There are 63 cross-border river basins, meaning that there is a general interdependence between countries concerning water courses. The most remarkable are those of the Nile, and of the Congo, Niger and Senegal rivers.

These connections between countries can be a source of tensions, as in the case of the Nile, which is the focus of a disagreement between Egypt and Ethiopia: the latter has started to build a hydroelectric dam that is likely to affect the countries lying downstream.

However, there are also many examples of cooperation. In the case of the Senegal River, several hydroelectric infrastructures are managed by a regional institution, the Senegal River Basin Development Authority (OMVS), for the benefit of all the countries in the region. Cooperation also proves to be necessary for managing the large lakes divided between several countries, the good ecological status of which is a major regional challenge: Lake Victoria, the world's second largest body of fresh water, is severely threatened, for example, by a risk of eutrophication, caused by excessive inputs of nitrogen and phosphorus, leading to a rise in plant biomass that makes life in the lake water more difficult.

On the oceans around the continent, fish migrations and also piracy represent common challenges that cross the maritime borders defined by the law of the sea.

CONTINENTAL COOPERATION TAKING FORM

Climate change is further reinforcing this need for cooperation and the countries of Africa are seeking to define rules for shared management. Water is the main vector of the risks and hazards associated with climate change—rising water levels, coastal erosion, more intense droughts and desertification, and flooding. In addition, despite its apparent abundance, at least 25 African countries could find themselves facing serious difficulties in sourcing potable water by 2025 (see pp. 80–81).

The "African Water Vision for 2025" adopted for the purpose of organizing "equitable and sustainable use of water for socio-economic development" recommends solidarity in the management of water resources between those countries that share the same hydrographic basins. Likewise, the 2050 Africa's Integrated Maritime Strategy (AIM Strategy) conceived by the African Union to endow the states with a common institutional response to the challenges of maritime security and the development of activities linked to water resources, promotes cooperation between stakeholders from the public and private sectors and from all levels of civil society. It also makes provision for the creation of regional operational centres in charge of ensuring maritime safety and security in their respective zones, working with their counterparts across the continent.

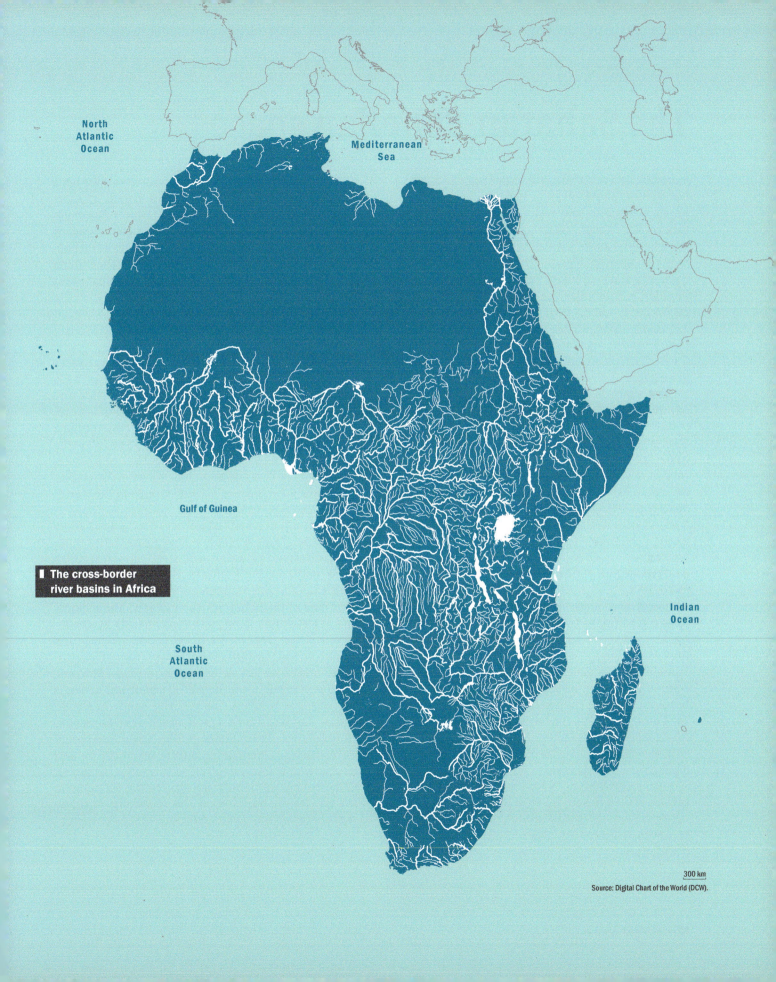

The Saharan-Sahelian space

A UNIQUE SPACE EXTENDING FROM THE SAHEL TO THE MEDITERRANEAN

Africa is home to the world's largest hot desert, the Sahara. The "Great Desert", as it is called in Arabic, stretches over more than 4,800 km from east to west, from the Atlantic Ocean to the Red Sea, and between 1,300 km and 1,900 km from north to south, from the Mediterranean to the Sahel-Sudan strip. It covers 8.6 million km², about one-third of the area of the continent, and stretches into the territory of 10 countries: Morocco, Algeria, Tunisia, Libya and Egypt along the northern edge, and Sudan, Chad, Niger, Mali and Mauritania along the southern edge.

The Sahara is therefore a cross-border space by nature, containing almost 17,000 km of borders—40% of the circumference of the Earth if put end to end!—and the management of which is a genuine collective challenge for the states in the region. It comprises vast areas with very few people and its population is estimated to be about 15 million, contrasting with the density of settlement along the coasts of the countries of North Africa and in the cities of the Sahel (see map).

A LITTLE-KNOWN SPACE OF MOVEMENTS AND ACTIVITIES

Although very sparsely populated, this desert has been a traditional place of exchanges for centuries. Ancient routes across the Sahara still remain—from Timbuktu to Fez and Marrakesh, or from Gao to Tunis, Tripoli or Cairo, for example. They facilitated commercial exchanges between the southern and northern parts of the continent and contributed to the emergence of the great African empires of West Africa, notably that of Gao. These activities declined with the colonial division of the Sahara and the rise of maritime routes between Africa and West Africa.

It was only from the 1970s onwards that the desert saw an upturn in activity, with the exploitation of its natural resources (hydrocarbons, ore) and the growth of towns triggered by sedentarization of nomadic populations.

Despite the lack of infrastructures (see map), the Sahara has once again become a region of transit, where there is trade in legal products (livestock, salt, etc.) and also illegal products (contraband goods, drugs and arms trafficking). The mobility of the population of the Sahel and Maghreb reflects the human and commercial ties between these two spaces via the Sahara.

COMPLEMENTARY FEATURES AND SHARED CHALLENGES

Currently, the Sahel is facing a major security crisis. The collapse of the Libyan regime in 2011 made supervision of the long borders more complex and has favoured movements of considerable quantities of armaments. The Mali crisis has been another factor triggering security tensions that were latent on account of long-standing conflicts between groups in the region. This situation has proved to be conducive to the emergence of jihadist groups in the region, thus increasing insecurity.

This situation has consequences for the lives of the local population—notably a sharp rise in the number of displaced persons—and for their food security. More particularly, it affects the economic, social, political and environmental challenges faced by the countries of the Sahel. For the moment, the regional and international responses have mainly been conceived at the level of the Sahel countries and their neighbours in the Gulf of Guinea.

The Saharan-Sahelian space could, however, become a venue for enhanced cooperation between the states of North Africa and the Sahel on shared issues. And there are many of them: developing agricultural and pastoralism, reinforcing cross-border infrastructures, managing natural resources (in particular water) and adapting to climate change. All these issues can become drivers for the economy in the Sahara, by building on the continuity it represents between the north and south of the continent.

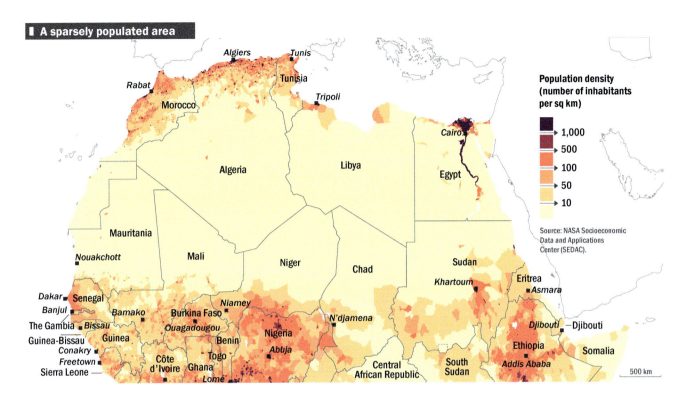
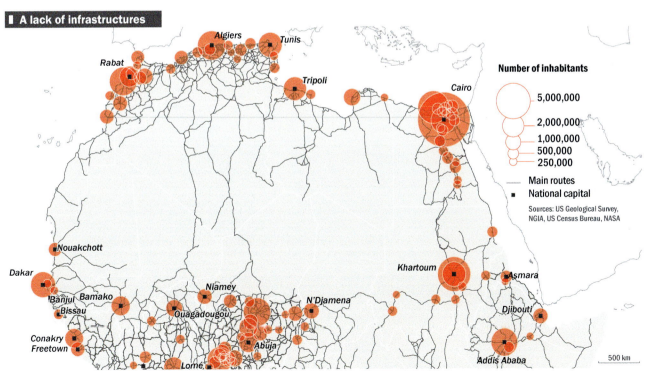

A MULTIFACETED CONTINENT WITH SHARED CHALLENGES

African forests under pressure

AFRICAN FORESTS: DIVERSE AND TRANSNATIONAL

Africa is home to about 16% of the total area of the world's forests. The continent's forests are highly varied. Their composition depends on combinations of biogeographic and climate factors that are specific to each region, and which have given rise to a wide variety of ecosystems, namely rainforests in the equatorial regions, open forests or grasslands in the tropical and sub-tropical savannah regions, Mediterranean forests in North and Southern Africa, and mangrove swamps in coastal wetlands, etc.

Some of these eco-regions host sub-ecosystems with specific local characteristics in just one country, such as the Knysa high-altitude forest in South Africa. However, some immense forests cover several countries, such as the forest in the Congo Basin and the Miombo Forest on the Zambezi, which stretch between six and seven countries, respectively. The biogeographical areas of these large forests generally cross over national borders, making them transnational shared spaces (see pp. 80–81).

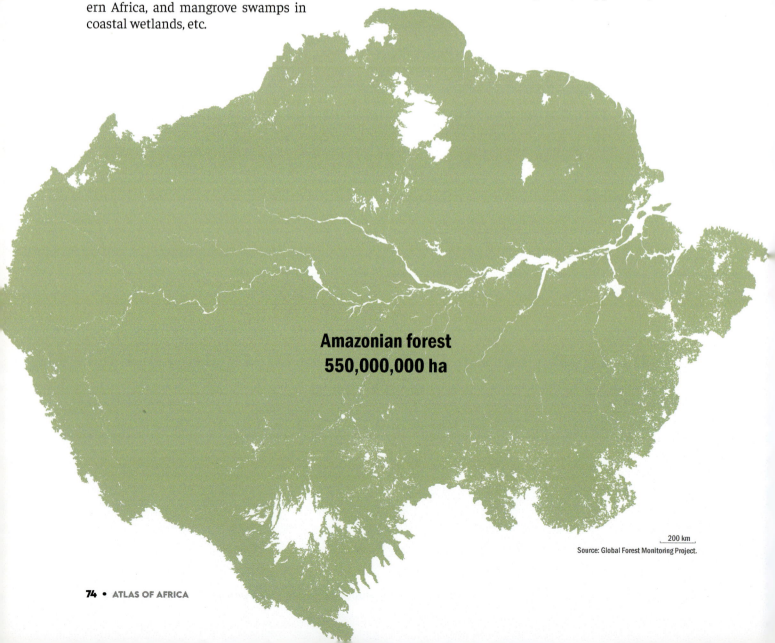

Amazonian forest 550,000,000 ha

Source: Global Forest Monitoring Project.

THE THREAT OF DEFORESTATION

Along with South America, the African continent has the highest rate of deforestation, with an average annual loss of 0.49% of its forest area between 1995 and 2015, compared to 0.13% at the global level. The problem appears to be getting worse: during the period 2000–2015 four of the eight countries that lost the greatest area of forests were African (Nigeria, the United Republic of Tanzania, Zimbabwe and the Democratic Republic of the Congo—DRC).

The pressures weighing on these ecosystems are primarily anthropic and are becoming more intense. The causes can be slash-and-burn farming, collection for firewood, or felling to build the transport infrastructures linked to urbanization and population growth. Forestry activities also play a role, directly for the industrial exploitation of wood and sometimes for trafficking of precious woods, but also to convert natural forests into profitable crops (rubberwood, palm oil, etc.).

The causes and intensity of the damage vary according to the type of forest, but their progression everywhere is bringing into question the multitude of services provided by forests: housing for people, natural resources, wildlife sanctuaries, carbon storage, etc.

THE FORESTS OF THE CONGO BASIN SYMBOLIZE THE CHALLENGE FOR AFRICA

With a land area in excess of 200 million hectares, the Congo Basin forest is the planet's second largest "lung", after Amazonia. As the latter suffers from an acceleration in deforestation driven by growth in the agricultural sector, it is feared that the largest forest in Africa may also face a comparable fate.

For the moment, deforestation in Africa has been contained to some extent, unlike in Amazonia; in the DRC, the area covered by forest shrank by 5% between 1990 and 2016, compared to a decline of 10% in Brazil. But the deterioration is still accelerating: in 2018, DRC posted the second-largest loss of primary forests after Brazil, and they could disappear completely by 2100. Development of roads, mining and uncontrolled felling are the main causes.

These multiple trends relating to forests are partly inevitable, but can be set right to some extent. That is the major conservation challenge for all the forests in Africa: to develop forest zone management and development policies that reconcile the needs of the population and the conservation of the essential functions of wooded spaces in the ecological equilibrium.

Congo Basin forest 268,000,000 ha

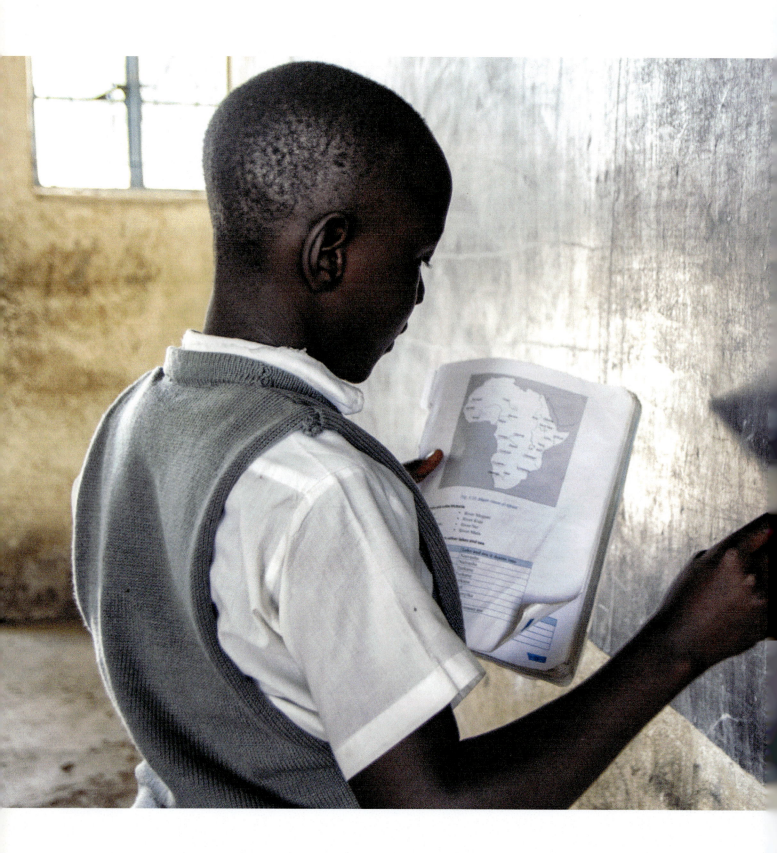

AFRICA REINVENTING ITSELF AND TAKING UP THE KEY CHALLENGES OF TOMORROW

Africa is changing fast and is reinventing itself along the way. In a fast-moving and often uncertain world, the continent is currently facing major challenges that will be decisive for its future. These challenges are diverse and complex (impacts of climate change, pressures on biodiversity, protecting human capital, governance issues, reinforcing social cohesion, etc.) and manifest themselves in a specific manner on the continent. Across the African continent, inventive and innovative responses to these local and global challenges are appearing. Stakeholders on the continent are seeking solutions beyond their own borders and imagining responses that can take account of the characteristics of Africa. This section of the Atlas therefore proposes some possible solutions to these key challenges that have been deciphered here, through just a few examples of the resources Africa has at its disposal to prepare for the future.

Primary school in Angira, western Kenya.
Photo © Lambert Coleman/AFD.

A continent that is vulnerable to climate change

THE AFRICAN CONTINENT POLLUTES THE LEAST…

The African continent emits relatively little in the way of greenhouse gases. Although they represented 16.6% of the world population in 2017, the African countries produced just 3.5% of the world's cumulative CO_2 emission during the period 1990–2014.

The fact that Africa makes a small contribution to air pollution can be explained by overall emission levels that are still limited in terms of energy consumption, industrialization or air and road traffic. Most of the emissions produced by the continent are the result of farming activities. In 2016, only three African countries with considerable mining and petroleum activities and particularly high-carbon energy production—South Africa (15th largest emitter globally), Zambia (16th) and Nigeria (18th)—were among the world's top 20 emitters. In that year, only 11 African countries emitted more than the global average of 6.76 metric tons of CO_2 per inhabitant.

… BUT IS THE MOST VULNERABLE TO CLIMATE CHANGE

However, the African continent cannot escape the effects of climate change, however. Admittedly, the rise in annual temperatures through to the end of the century, projected at 1 to 4 degrees Celsius in Africa depending on the different emissions scenarios taken by the Intergovernmental Panel on Climate Change (IPCC) and the World Bank, will be slightly less marked than that expected worldwide (between 3 and 6 degrees Celsius). However, as the average temperature is already high on the continent, North and Southern Africa are likely to be hit by exceptional heatwaves. These should also turn out to be more frequent and longer in the tropical zones than in the rest of the world.

The effects of global warming could be accentuated by the weak resilience and adaptation capacities of certain countries. According to the ND-GAIN Index that ranks these two dimensions of exposure and readiness on a scale of 100, Africa has a much lower score (40) than the global average (52). One indication of this vulnerability is that cereal yields could fall by 5% to 24%, depending on the crop in question and the scenario.

These trends could cause population displacements within the continent: population density could leap by 300% through to the end of the century in the sub-tropical areas which are cooler than the equatorial and tropical zones.

FINANCE AT THE SERVICE OF CLIMATE PROJECTS: THE EXAMPLE OF THE GREEN CLIMATE FUND

The United Nations Development Programme (UNDP) estimates that Africa will need $50–$100 billion per year to take up the climate challenge through to 2050. These funds should allow the financing of greenhouse gas emission reduction projects, and also of projects to facilitate the adaptation of local territories and their population to the upcoming changes.

The Green Climate Fund (GCF) was set up in 2011 and constitutes a response to this issue. Initiated by the United Nations Framework Convention on Climate Change (UNFCCC), the GCF is an instrument for financing climate projects in developing countries, and more particularly in Africa. Some 49 projects (valued at $2.2 billion) have already been financed in the areas of renewable energy production, flood management, resilient agriculture, clean cooking and the restoration of ecosystems.

This pioneering facility is the perfect example of the financial mechanisms that are required to support the continent's efforts to prepare itself more effectively for the coming upheavals (see pp. 110–111).

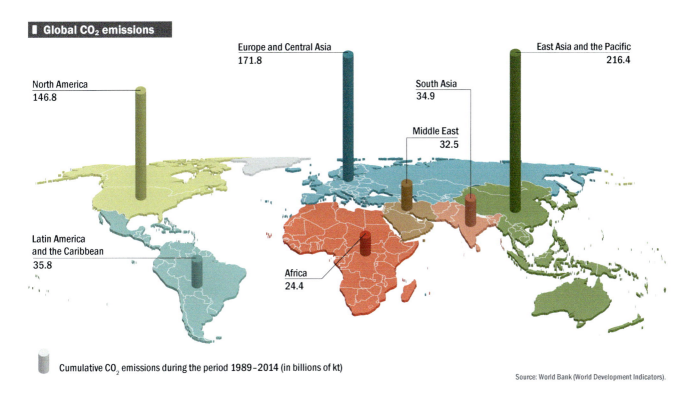

Global CO₂ emissions

- North America: 146.8
- Latin America and the Caribbean: 35.8
- Europe and Central Asia: 171.8
- Africa: 24.4
- Middle East: 32.5
- South Asia: 34.9
- East Asia and the Pacific: 216.4

Cumulative CO_2 emissions during the period 1989–2014 (in billions of kt)

Source: World Bank (World Development Indicators).

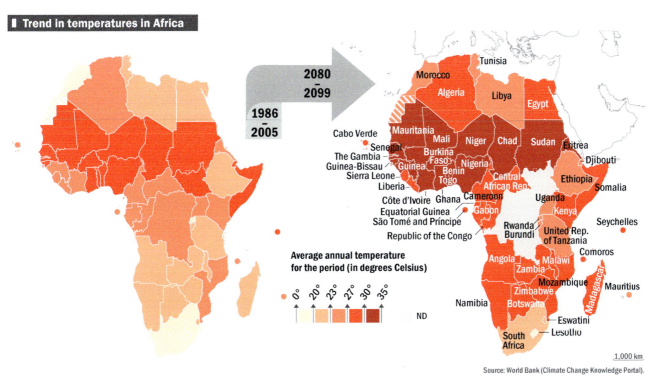

Trend in temperatures in Africa

1986–2005 → 2080–2099

Average annual temperature for the period (in degrees Celsius): 0°, 20°, 23°, 27°, 30°, 35°, ND

Source: World Bank (Climate Change Knowledge Portal).

AFRICA REINVENTING ITSELF AND TAKING UP THE KEY CHALLENGES OF TOMORROW • 79

The impacts of climate change

EITHER TOO MUCH OR TOO LITTLE: CLIMATE CHANGE, WATER RESOURCES AND RAINFALL

Due to climate change, the continent will find itself facing more frequent extreme climate events, such as flooding or droughts: while devastating cyclones are hitting millions of people in Mozambique, Malawi and Zimbabwe, other areas in North Africa and far south of Southern Africa are facing increased risks of water stress (as drinking water needs exceed available resources). The city of Cape Town in South Africa has thus been placed under the constant threat of its first "day zero", which would require the water supply to all its inhabitants to be cut off.

The mechanisms at play are often complex, however: in certain parts of the Sahel, the relative rise in rainfall levels compared to the 1990s is going hand in hand with more intense rain events which lead to flooding. Faced with this growing uncertainty, the immediate challenge is to develop resources such as meteorological and climate information systems to anticipate changes more effectively and adapt to them.

THE AFRICAN COASTLINE UNDER THREAT FROM RISING SEA LEVELS

The rising sea levels along the African coastline are another very immediate challenge induced by climate change. Sea levels could increase by more than one metre by the end of the century, according to the Intergovernmental Panel on Climate Change (IPCC). Yet these areas are home to some of Africa's largest cities, which often feature a large proportion of the economic activity and population (see pp. 52–53).

Three African cities (Alexandria in Egypt, Lagos in Nigeria and Abidjan in Côte d'Ivoire) are thus among the 20 cities in the world that are the most endangered by rising sea levels. This trend could expose more than 10 million African city-dwellers to direct risks, such as submersion of their living quarters, unavailability of drinking water, or damage to their urban infrastructures. The material damage could be particularly significant for certain coastal countries: in the least extreme scenario which corresponds to a 1 metre rise in sea levels, the minimum cost of the damage caused by the rise in sea levels is therefore estimated at 9.4% of GDP in Mauritania, 6.4% of GDP in Egypt and 5.6% of GDP in Benin.

BUILDING AFRICAN CITIES SO THAT THEY ARE RESILIENT TO CLIMATE CHANGE

While rural areas and agriculture in Africa are particularly vulnerable to the effects of climate change, the current urbanization process also requires us to anticipate the consequences for the cities. The most densely populated urban areas will no doubt suffer the biggest impacts in terms of health and energy consumption. Urban planning is central to the response and must take account today of the risks relating to rising sea levels, flooding and heatwaves in cities.

The fight against flooding may imply the building of infrastructures such as effective drainage systems along roads. In coastal zones, simple constructions such as dykes may provide short- or medium-term protection for districts to allow time to transfer flood waters gradually elsewhere. In the city, the creation or restoration of green spaces in areas subject to flooding features among the solutions to cope with spells of intense heat, and also to reduce flooding during the rainy season.

Finally, the development of early warning systems for extreme climate events, the reinforcement of civil protection services and the development of dedicated financial tools can facilitate better preparation for when climate disasters do occur.

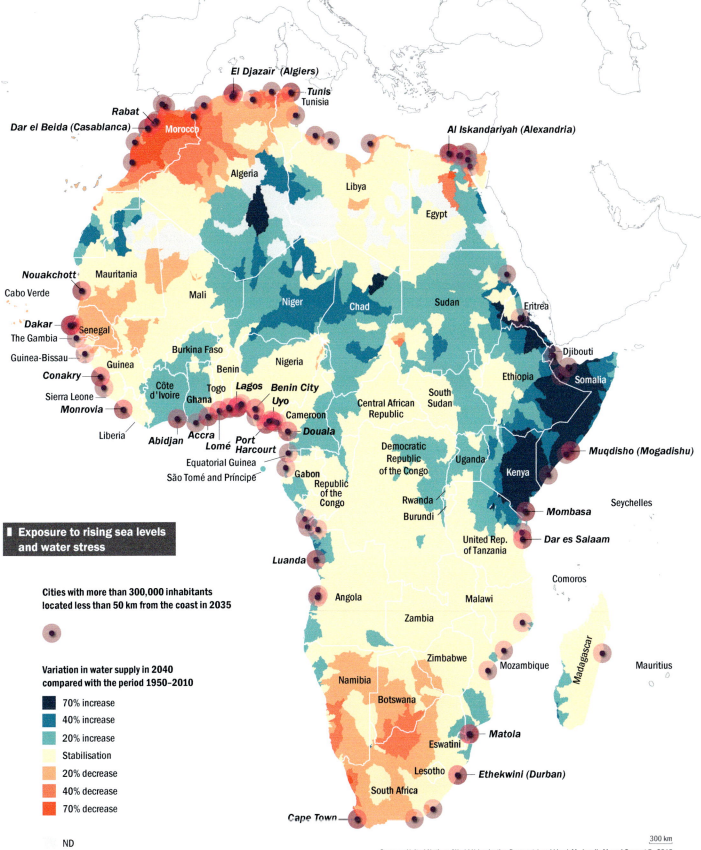

African biodiversity under pressure

IMMENSE BUT ENDANGERED BIODIVERSITY

Africa is home to almost one-quarter of all the living organisms on the planet and, according to the International Union for Conservation of Nature (IUCN), some 14% of the known flora and fauna that can be found in only a single country - referred to as endemic. However, this remarkable biodiversity is endangered. It is estimated that 42% of the endemic species in Africa are threatened with extinction, which represents almost 850 animal and plant species, to which should also be added the non-endemic species (see map).

All the different habitats are under pressure—forests, savannah, wetlands, oceans, etc.—and this affects mammals, birds, reptiles, plants and fish indiscriminately. The most vulnerable specimens include some of the most emblematic animals to be found on the continent, such as elephants, lions and giraffes.

The sharp decline in the numbers of the latter species, by almost 40% in 30 years, led the international community to place the giraffe on the red list of species at risk of extinction, in 2016.

Protected spaces around the world (2016)

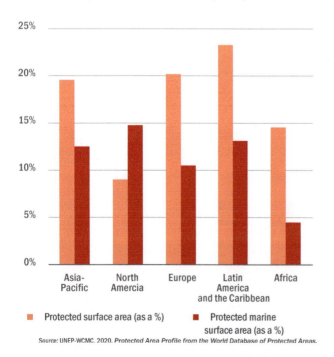

Source: UNEP-WCMC, 2020, Protected Area Profile from the World Database of Protected Areas.

ECOSYSTEMS THAT ARE NECESSARY FOR MANKIND

In addition to their function as habitats for living organisms, these natural spaces provide what are known as "eco-systemic" services to mankind: provision of resources (food, water, firewood, etc.), socio-cultural services (landscape, tourism, etc.) and environmental regulation services (air and water quality, soil fertility, etc.). In addition, the deterioration of ecosystems has an impact on climate change, as one of their functions is to capture carbon and that carbon is released into the atmosphere as they shrink.

Wetlands (mangroves, marshlands, seagrass areas), of which the continent is home to more than one-fifth of the total world surface area, combine all these different factors. These ecosystems play a fundamental role for the coastal population, by providing them, among other things, with food (fish, rice), economic activity (fisheries, tourism) and protection against coastal erosion. They also form one of the main carbon sinks on the continent, with storage capacities exceeding those of the tropical forests. Owing to coastal development, however, this natural capital is deteriorating rapidly, and it is thought to have lost 25% of its surface area since the 2000s.

PROMOTING THE EXTENSION AND FUNCTIONING OF PROTECTED AREAS

The protection of natural habitats is one of the solutions to preserve biodiversity in Africa. In 2019, the continent had almost 8,500 areas protected under a variety of local systems (nature reserves, national parks) and international rules (413 Ramsar-protected wetland sites, 48 UNESCO natural and mixed heritage sites, etc.) covering a total of about 15% of the territory of the continent, and 5% of its maritime space.

These areas could be extended further: on the other continents, they cover an average of 18% of the land surface and between 10% and 15% of the marine area (see map).

These protected areas are much more than just a status: they are a method of management that should be promoted. In Mauritania, the Banc d'Arguin

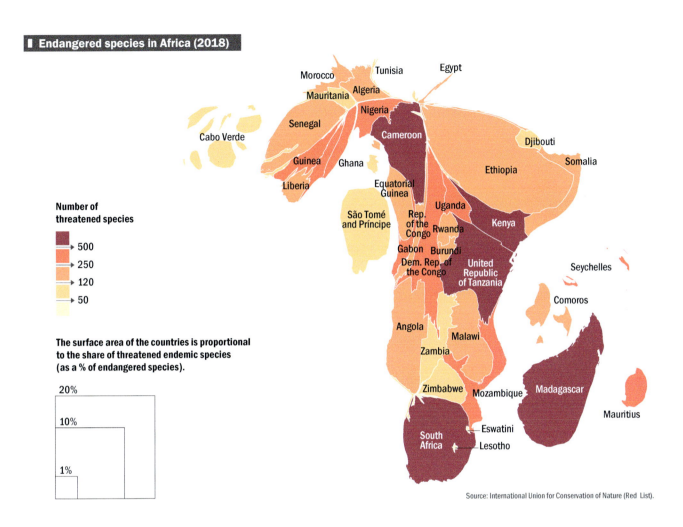

Reserve, a marine area that is essential for biodiversity not only in the region (fishing resources), but also internationally (migratory birds), is under threat from different forms of pollution and from over-fishing. Since 2012, the Endowment Fund for the Banc d'Arguin and Coastal and Marine Biodiversity in Mauritania has been financing activities in various areas, such as surveillance, community development, ecological monitoring and environmental education.

This innovative funding mechanism thus contributes to maintaining the services rendered by these ecosystems, the annual value of which has been estimated at €198.8 million per year.

Skills acquisition and enhancement: the education challenge in Africa

COPING WITH THE LEARNING "CRISIS"

Despite the progress made in terms of school enrolment among children in Africa (see pp. 12–13), the education sector is facing an issue of quality. The gaps in today's education systems are limiting the ability of school institutions to allow children and teenagers to reach the minimum skill levels required in reading and mathematics. For example, 88% of children and teenagers are not capable of reading fluently when they have completed primary and lower secondary education in Africa (see map). The continent also suffers from a phenomenon of children dropping out after primary schooling.

The completion rate, meaning the percentage of pupils in a given generation who are enrolled in the first year of a level of teaching and considered as having completed it, which stands at 75.4% in primary schooling, falls to 50.1% for lower secondary education and to 25.3% for upper secondary level (high school) (see map). Access to higher education, meanwhile, is restricted to an extremely limited number of African citizens, with a completion rate of less than 8.1%. These difficulties in access to education are all the greater for the poorest people, girls and people living in rural areas.

ACTING ON THE QUALITY OF TEACHING CONDITIONS

In many countries, schools do not provide conditions conducive for learning and instruction, on account of the poor resources at their disposal. First, there are the material issues: when they are at school, a large proportion of African children learn without any teaching materials (schoolbooks) and have access neither to potable water nor to toilets. In addition, the teachers are often too few in number and are under-qualified (see graph). In primary education, for example, class sizes are too large, with a pupil-to-teacher ratio of 38:1 (down from 43:1 at the start of the 2000s) and the teachers do not have sufficient training.

Africa is the region facing by far the biggest challenges and accounts for two-thirds (67%) of the additional teachers required to achieve universal primary education by 2030.

INVESTING IN EDUCATION: IMPROVING THE TEACHING OF SCIENCE AND TECHNOLOGY

Another important challenge for Africa's education systems is improving the teaching of science and technology, in order to contribute to the economic and social transformation of the continent.

Many African countries are making a commitment to improving teaching in this area, such as Rwanda which dedicates 14% of its education budget to financing with a focus on STEM teaching (science, technology, engineering and mathematics). Developing these skills requires investment through the

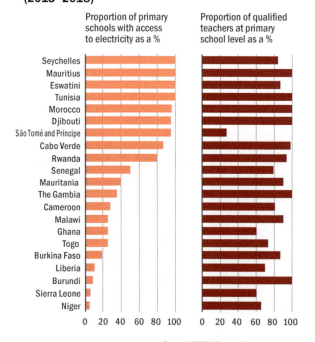

Lack of equipment and teacher training in primary schools (2013–2018)

Source: UNESCO (Sustainable Development Goals).

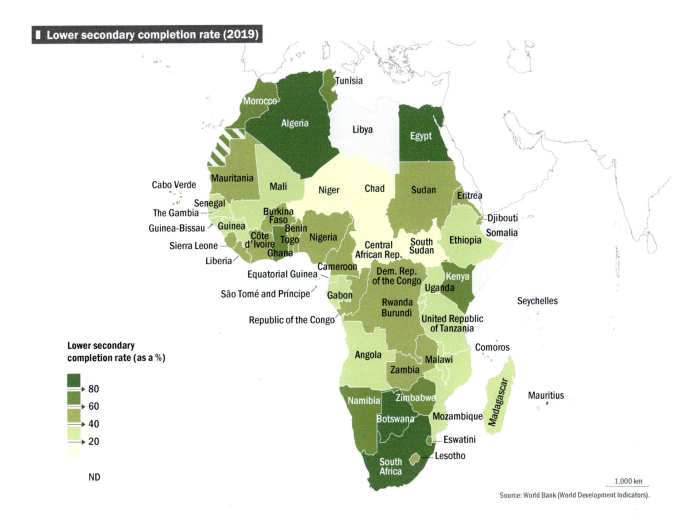

Lower secondary completion rate (2019)

Source: World Bank (World Development Indicators).

whole continuum from early learning through to higher education and research.

In-service training for teachers in experimental methods, such as the system developed by the "La Main à la Pâte" Foundation, is a key driver contributing to improving the quality of teaching. In Senegal, the ADEM project to support the development of lower secondary teaching is already rolling out a systemic approach combining an improvement in school capacities and conditions (furnishings, teaching materials and IT) and the skill levels of pupils (remedial teaching for children struggling at school) and teachers (training in investigation methods). Another example is the creation of 58 centres of excellence across 19 African countries, to foster the development of scientific research in Africa in the areas of science and technology.

Employment in all its forms: a key to the prosperity of the continent

MASSIVE YOUTH EMPLOYMENT NEEDS

Africa is the region which is set to record the biggest rise in its active population (aged 15–64 years). According to United Nations forecasts, by 2050 it will represent over 75% of growth in the global workforce. The International Monetary Fund (IMF) and the World Bank thus estimate that there are almost 440 million young people who will join the African labour market by 2035 (see diagram), which is the equivalent of the combined current population of the United States and Mexico. The capacity of the African economies to create the necessary jobs and that of the new arrivals to find their place in the labour market are therefore issues of key importance. In this respect, there are marked differences between sub-regions today.

In Southern and North Africa, the unemployment rate among young people aged 15–24 is high (26.2% and 30.6%, respectively). The countries of West, Central and East Africa record the lowest official youth unemployment rates (11% on average). This figure does not account for the extent of under-employment in these regions, however, which is more difficult to quantify.

EMPLOYMENT QUALITY AND INFORMAL EMPLOYMENT

The benefits that Africa can gain from the rapid increase in its active population will depend not only on the number of jobs created, but also on their "quality". This challenge of decent job creation is made all the more complex by the largely informal nature of the jobs that currently exist on the continent, offering little in the way of security to workers. With the exception of the most advanced countries of North and Southern Africa, almost 90% of jobs fall outside the formal framework in a large majority of the countries on the continent (see diagram). This informal sector encompasses jobs requiring different qualification levels (craftspeople, taxi drivers, fishermen, etc.) and may also concern companies numbering several dozen people. One common feature of these informal jobs, however, is the weakness of the legal, social and health protection they provide. The effective unemployment insurance coverage rate in Africa is less than 1%, compared to 4.6% in Latin America and 7.2% in Asia.

THE ROLE OF POLICIES IN SUPPORTING EMPLOYMENT AND ENTREPRENEURSHIP

In this context, entrepreneurship is no doubt set to play a specific role in job creation in Africa. The African population is already the most entrepreneurial in the world: 22% of Africans of working age create new companies, compared to 19% in Latin America and 13% in Asia. If they do so, it is largely owing to the lack of prospects in salaried employment.

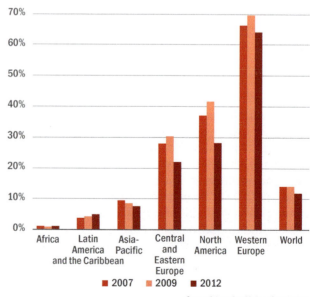

■ Effective unemployment insurance coverage

Source: International Labour Organization.

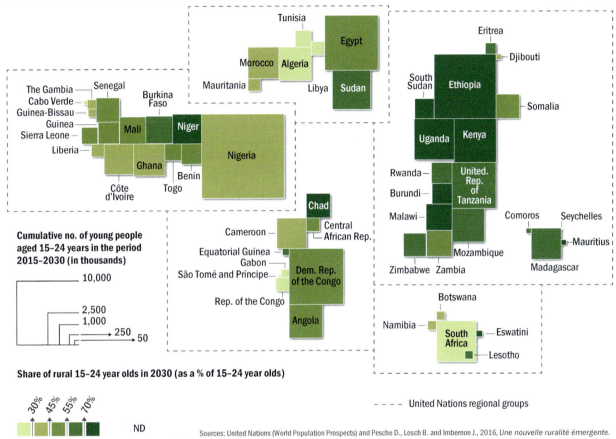

Number of young people joining the labour market (2015–2030)

Sources: United Nations (World Population Prospects) and Pesche D., Losch B. and Imbernon J., 2016, *Une nouvelle ruralité émergente. Regards croisés sur les transformations rurales africaines*, Atlas pour le programme Rural Futures du NEPAD, Montpellier, Cirad-NEPAD.

In the absence of any alternatives, many Africans start their own businesses in the form of self-employment, family-run companies and/or micro-enterprises, in a wide variety of areas ranging from agriculture to commerce, and including crafts activities and transportation. And there is immense further potential for promoting entrepreneurial culture in Africa to generate jobs and economic autonomy for the population.

In order to support entrepreneurs, certain public and private sector are actively devising and rolling out ambitious and innovative policies. In Chad, for instance, the Maison de la Petite Entreprise (Small Business Centre) in N'Djamena delivers customized services to micro-entrepreneurs seeking to create or consolidate their independent business activities. The centre conducts awareness raising and hosts, guides, trains and advises young business owners during the different stages in the creation or management of small and medium-sized enterprises. For the more advanced, serious projects, it also proposes guidance in gaining access to financing.

Issues concerning women's status in Africa

WOMEN AS ACTORS IN THE SUSTAINABLE DEVELOPMENT OF THE CONTINENT

African women are heavily involved in economic life. In the continent as a whole, their workforce participation rate stands at 55%, which is higher than that observed in the most advanced countries (53%) and than the world average (48%) (see map).

However, this overall situation does hide some disparities within the continent. In North Africa, participation of women in the labour market is among the lowest in the world, at 21% on average, compared to 71% for men. The women in this region are almost three times as likely to be affected by unemployment as men.

Another indication of the key role played by women in the economy is that many of them start up their own businesses. According to a study conducted by Women in Africa Philanthropy, 24% of African women have set up their own companies, far more than their sisters in Latin America (17%), North America (12%), South-East Asia (11%) or the Middle East (9%).

INEQUALITY BETWEEN THE SEXES AND THE CHALLENGE OF FEMALE GENITAL MUTILATION

However, this particular role of women in the African economy does not prevent the existence of inequalities, as in other regions of the world, between men and women, in a variety of areas: education, health, economic opportunities or participation in political life According to the World Economic Forum, the scale of these inequalities in sub-Saharan Africa can be considered very much similar to that observed in South-East Asia, with the situation in North Africa standing out as being more negative.

However, situations do vary from one country to another—Rwanda ranks among the 10 most egalitarian countries in the world—or from one context to another; for instance, rape is used as a weapon of war in regions where there are conflicts. Inequalities between men and women in sub-Saharan Africa have actually declined in certain areas, and notably in education.

However, certain regions of Africa have attracted negative attention globally owing to female circumcision and genital mutilation (see map). Currently, 91.5 million women and girls over the age of nine years are living with the consequences of sexual mutilation. These practices are particularly widespread in Egypt and in certain countries of the Horn of Africa (Somalia, Djibouti, Eritrea) and West Africa (Guinea, Sierra Leone, Mali), where more than four in five women are affected.

INITIATIVES IN FAVOUR OF WOMEN TO BUILD A BETTER FUTURE

There are a certain number of concrete responses to this particular societal issue. Supporting female entrepreneurship, for example, provides an opportunity to promote both gender equality and local development. In Egypt, a support programme for small companies headed by women is designed to facilitate access for these female entrepreneurs to financing and to recognition of their potential by private actors and institutions—two things that often constitute major obstacles for them. Their success is a demonstration that should consolidate the growing awareness that women, who are often given a domestic role in Egypt, can also make their contribution to transforming the country.

In situations of conflict, women who suffer sexual violence can also receive support. This is what is proposed by a specialized centre in Bangui, in the Central African Republic, which has been suffering from a violent conflict since 2013. The centre draws its inspiration from Panzi Hospital run by Denis Mukwege, winner of the Nobel Peace Prize in 2018, in the Democratic Republic of the Congo, and provides a medical, psychological, legal and socio-economic response to women who are victims of violence in the country.

Obesity and chronic diseases: new health challenges

THE PROGRESSION OF NEW PATHOLOGIES

To varying degrees according to their level of development, all the countries of Africa are facing a progression in mortality from diseases that used to be rare on the continent: chronic diseases such as diabetes, cancer or cardiovascular diseases, etc., as well as neurodegenerative diseases, such as Alzheimer's.

Ageing of the population and changes in lifestyles driven by urbanization are the main causes. Consumption of alcohol, sugars and animal fats, in particular in processed products, gives rise to risks that have worsened with the decline in physical activity, and with smoking and exposure to pollution.

These new public health challenges are already making themselves felt in Africa. Pathologies linked directly to changes in lifestyles, such as obesity, are increasingly widespread (see map). Diabetes is thought to affect 6.3% of African adults today, compared to 4.8% in 2010. The diabetes prevalence rate is particularly high in North Africa, where it stands at 12.3% (see graph). While the leading causes of mortality are still respiratory infections, HIV and diarrhoea, split cardio-vascular diseases are now catching up with them.

AFRICA'S EPIDEMIOLOGICAL TRANSITION

Although Africa is showing a marked reduction in the prevalence of infectious, parasitic and deficiency-related diseases, the appearance of these new health issues marks the beginning of an epidemiological transition. This is characterized by a fall in mortality due to transmissible diseases and an expansion in chronic and degenerative diseases. It is the transition that accompanies the development process, reflecting improved hygiene, nutrition and access to healthcare in the population in question on the one hand, and their confrontation with "new" pathologies, on the other.

These diseases are particularly poorly diagnosed in Africa today. This is true in almost 62% of diabetes cases on the continent (see graph).

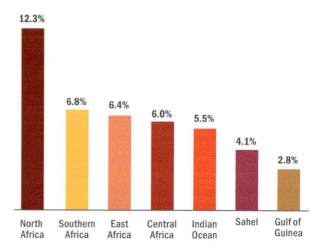

Prevalence of diabetes among adults by region (2019)

As a % of the population aged 20–79 years

- North Africa: 12.3%
- Southern Africa: 6.8%
- East Africa: 6.4%
- Central Africa: 6.0%
- Indian Ocean: 5.5%
- Sahel: 4.1%
- Gulf of Guinea: 2.8%

Source: World Bank (World Development Indicators).

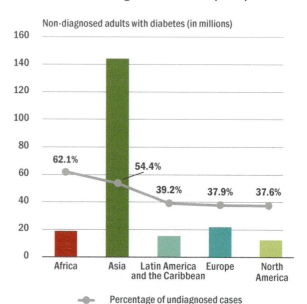

Persons with undiagnosed diabetes (2017)

Non-diagnosed adults with diabetes (in millions)

- Africa: 62.1%
- Asia: 54.4%
- Latin America and the Caribbean: 39.2%
- Europe: 37.9%
- North America: 37.6%

Percentage of undiagnosed cases

Source: International Diabetes Federation, 2017, *IDF Diabetes Atlas*, 8th edition.

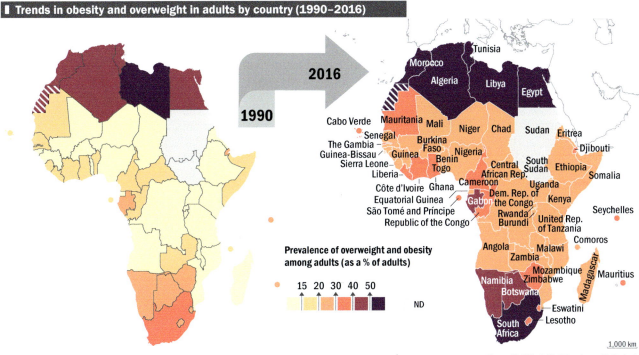

Trends in obesity and overweight in adults by country (1990–2016)

Prevalence of overweight and obesity among adults (as a % of adults): 15 20 30 40 50 ND

Source: World Bank (World Development Indicators).

REINFORCED HEALTH SYSTEMS TO TAKE UP THE DUAL CHALLENGE OF PANDEMICS AND CHRONIC DISEASES.

The epidemiological transition is still in its infancy in Africa, but the demographic trends on the continent are giving rise to a twofold challenge: to respond to the continuing problems of infectious diseases, which can only be eradicated gradually, and to that of the typical pathologies of the advanced societies, which are already emerging. This trend amplifies the need for efficient health services, which are necessary for the prevention, detection and treatment of these numerous pathologies, some of which remain relatively unknown locally. It thus constitutes an economic challenge for societies, as the handling of these new diseases will lead to considerable increases in the health expenditure incurred by states and households.

This dual challenge requires the development of a more integrated approach in order to reinforce health systems in a lasting manner by addressing all their components (infrastructures, equipment, medicinal products, human resources and governance). The health sector is under-financed in many countries and is highly reliant on international aid, but this tends to be directed more towards programmes targeting a disease or group of diseases, an approach that needs to be changed to reinforce care systems as a whole.

High expectations for the rule of law and freedom of expression

DEMAND FOR JUSTICE

The trend towards democratization that is underway in Africa at the political level is driving increased grass-roots demand for the establishment of the rule of law and for more justice. Africans willingly accept the principles of the judicial system—74% of them think that judicial decisions should be abided by—but far fewer of them place trust in the courts. Surveys of the population do reveal that over 40% of the parties in judicial proceedings do not trust the courts, and one-third of the population considers judges to be corrupt or corruptible (see graph). In the course of the past decade, trust in the courts has tended to decline in Africa on the whole.

In the face of this, global efforts in favour of justice were stepped up in 2019 with the signing of the Hague Declaration on Equal Access to Justice for All by 2030 by representatives of countries and international organizations.

FREEDOM OF THE PRESS IN AFRICA: DIVERSE SITUATIONS

The press and news media (radio, TV, written press) are another pillar of the rule of law, and the ability of journalists to conduct their profession freely varies considerably from one region to another. The Reporters Without Borders (RWB) ranking, which conducts an annual assessment of media freedom and the resources implemented by states to protect that freedom, ranks Africa in fourth position, behind Europe, North America and Latin America, but ahead of Asia.

However, freedom of information takes a variety of forms on the continent. On the whole, it is better respected in West and Southern Africa, where several countries now have a situation qualified by RWB as "quite good", with a more flexible system of press freedom.

However, freedom of the press remains very limited in North Africa, Central Africa and East Africa (see map). In addition to this, the quality of the various media and their regulation are a challenge in many countries, accentuated by the growing role of social media as a vector of information facilitated by the development of the internet and mobile telephones.

INNOVATIVE INITIATIVES TO PROMOTE GRASS-ROOTS PARTICIPATION IN THE MEDIA

Demand for the rule of law is part of a growing insistence among Africans for a more participatory, inclusive form of democracy. In this respect, the media can play a key role in grass-roots dialogue, notably by providing a forum for groups whose voices are often the least heard, especially young people and women.

In Senegal, the "Bruits de Tambours" project has adopted an innovative approach in this respect, in order to change attitudes in West Africa. The project is based on the principles of education via entertainment through the media and consists of a communication campaign aiming to promote the participation of young people and women in public life.

By combining the use of mass media (a political thriller-style TV series, a radio soap opera in local languages), social media activities and local community action, the project intends to elicit questions and debate on individual behaviour and social norms such as sexism and gender inequality, breaking down preconceived ideas and developing an interest in dialogue and collective action among participants in the project. The storylines of the TV series and radio soap opera are devised in collaboration with the authorities and specialist partner associations.

Freedom of the press around the world (2019)

Freedom of the press
- Very serious situation
- Difficult situation
- Sensitive problems
- Quite good situation
- Good situation
- ND

Source: Reporters Without Borders.

Access to justice: perceptions and experience

Among all respondents

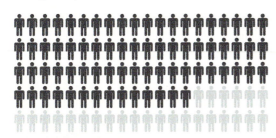

74.1% Agree with the statement that the courts have the right to make decisions that people always have to abide by.

42.5%
Have no or little confidence in the courts.

32.8%
Think that most or all judges are corrupt

13.3%
Have had contacts with the courts in the past five years.

Source: Afrobarometer.

Inventive civil society in Africa

THE VITALITY OF CIVIL SOCIETY IN AFRICA

Civil society, encompassing non-governmental organizations (NGOs), foundations, professional associations, trades unions or cooperatives, has developed in Africa since the 1990s. The trend towards democratization (see pp. 32–33) has gone hand in hand with the appearance of new forms of mobilization in favour of promoting human rights, and also economic and social rights.

Surveys conducted by the International Institute for Democracy and Electoral Assistance (IDEA) at the global level indicate that grass-roots engagement via civil society organizations has been growing in Africa since the 1990s, and has maintained its progression since 2010, despite the fact that it has been declining in the other regions of the world (see graph).

This dynamic in civil society is all the more remarkable in that it has been facing more severe restrictions on the activity of NGOs on the continent. These restrictions are evidence of the complex relationship between states and the different strands of civil society.

THE DIFFERENT FACES OF CIVIL SOCIETY

African civil society is highly diverse, both in terms of its modes and areas of action. Although promoting democracy is often central to this action, civil society in Africa is instigating and developing initiatives in areas such as education, employment, environmental issues, welfare protection and women's rights. These organizations may have different relationships with state structures and international cooperation actors, and may or may not interact with them. The diversity of civil society organizations, and their position close to the population, make them central to development issues.

ALTERNATIVE FORMS OF GOVERNANCE

Civil society in Africa is thus initiating and co-producing new forms of governance, acting at the regional, national and local level alike. Two examples testify to this diversity as a driver of change.

In 2012, in the wake of the Arab Spring, the Al Bawsala organization was created in Tunisia from a community of bloggers in order to establish a political observatory aiming to make citizens the central focus of public action.

The agenda of the organization takes the form of three major projects: (1) the parliamentary observatory; (2) the municipal observatory; and (3) the state budget and government action observatory. Its activities notably concern the collection and dissemination of information, advocacy efforts to promote

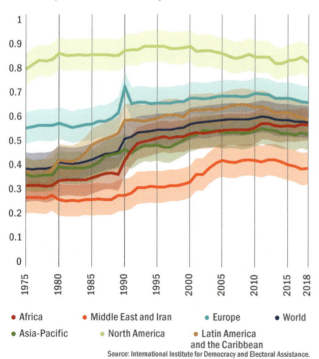

Participation in civil society

Source: International Institute for Democracy and Electoral Assistance.

The measurement of participation in civil society is based on three indicators: (1) the description of people's participation in civil society organizations; (2) the extent and independence of public debate; and (3) the frequency with which the main civil society organizations are consulted by political decision-makers. It is measured between 0 and 1 (strong participation in civil society).

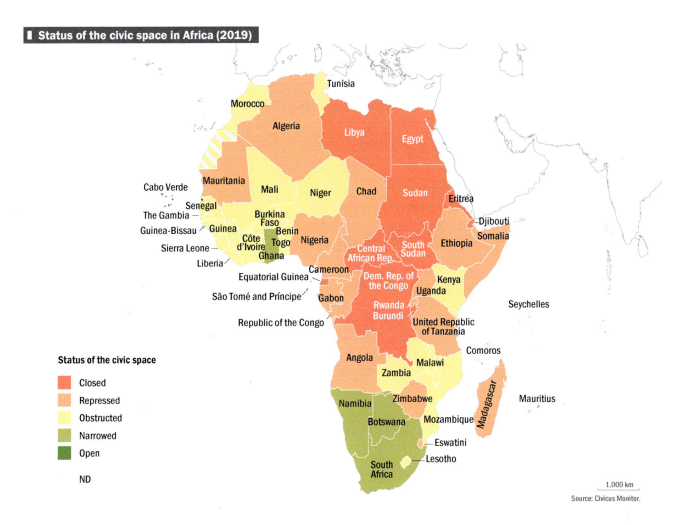

Status of the civic space in Africa (2019)

The civic space is defined as respect for freedoms of association, of peaceful assembly and of expression, by the law and in practice.

Source: Civicus Monitor.

transparency and the defence of fundamental rights, the drafting of bulletins, and the production of summarizations of parliamentary debates. In Niger, the "Les Puits du Désert" association works in the north of the country, a region where just 3% of the population have access to an improved water point and only 5% to potable water; near their camps, children (and in particular small girls) must walk for miles to find water, to the detriment of their schooling. The association is acting to improve the situation of the population by developing access to potable water and creating income-generating activities through the construction of wells and the creation of vegetable gardens.

Sport as a vector of economic and social development

SPORT AS A FACTOR IN ECONOMIC DEVELOPMENT IN AFRICA

Can sport contribute to economic development in Africa? As an economic sector, it does represent 2% of global GDP, providing jobs and generating economic activity through fees, licences, purchases of sports equipment, construction of sports infrastructures and all the "sport business" revolving around sports events (TV broadcasting rights, advertising, ticketing, etc.). In Africa, the sports sector remains little known and we have few figures. Although the "sports" expenditure of households will no doubt increase as purchasing power rises, the countries on the continent are counting more on the organization of sports events.

In recent years, a large number of countries have organized events on an international scale: the Africa Cup of Nations (football), the Francophone Games (multisports), the UCI Africa Tour (cycling), etc. The economic challenges relating to sport are considerable, however. Although the benefits of such major events are particularly difficult to measure—jobs, tourism, exports, international image of the country—the experience of South Africa in organizing the football World Cup in 2010 emphasized the challenges of coping with infrastructure construction costs when the budget situation is tight. The foreign successes of sporting talents have revealed the quality of African athletes, but this does also regularly raise the question of meeting the cost of their training.

Actors in the sports sector

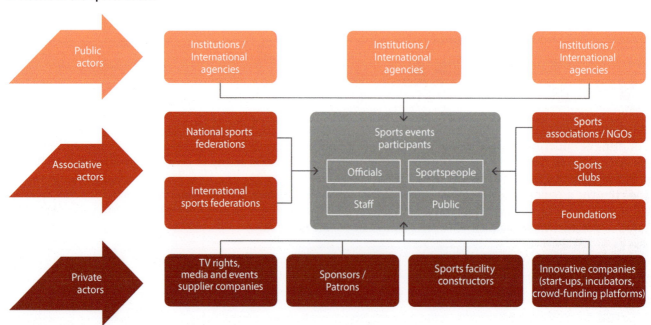

Source: AFD.

SPORT NEEDS TO BE STRUCTURED AS A DRIVER OF SOCIAL INCLUSION

Aside from its sole economic impact, sport appears to be a key area for a young continent (see pp. 6–7). It can contribute to personal fulfilment and the reinforcement of social ties. As a means of dialogue that can contribute to peace and reconciliation in contexts of ethnic or religious conflicts or tensions, it can also constitute a vector of social inclusion, of prevention in health matters and of education. However, in order to draw all the (economic and social) benefits from sport, the structuring of the sector needs to be enhanced. The development of sport implies having sectors that involve a number of public, non-profit and private sector stakeholders. It is through their cooperation that it becomes possible to provide adequate infrastructures and facilities, efficient institutional supervision, and professional development for the actors in the sector. In certain African countries, such as Morocco, Côte d'Ivoire or Kenya, the implementation of a sports policy is a key focus and the non-profit and private actors in sport are being reinforced.

SPORT AS A WAY TO LIVE TOGETHER BETTER: THE EXAMPLE OF BURUNDI

In Burundi, where one in two children does not finish primary school, early school leaving is a genuine problem. Playdagogy, a participative and fun method developed by the PLAY International NGO, is designed to detect children in the process of dropping out of school and encourage them to remain or return there. The NGO organizes educational and socio-sports projects in which 20,000 children aged between 5 and 12 years have taken part since 2016, in schools and community centres in 15 locations. Its actions encourage people to live together in harmony and combat all forms of discrimination, which is an essential component of any project in times of crisis. A network of over 190 Playdagogy practitioners has been set up in the country and 89 activity venues have been secured. In order to build on the success of the project in Burundi, the PLAY International NGO has launched a new, larger-scale Playdagogy project in Burundi and also in two other African countries, Senegal and Liberia.

The Playdagogy Programme of the PLAY International NGO designed to give children access to and keep them in school. Burundi, 2018.

Photo © PLAY International.

Solar energy in Africa: reconciling sustainability and economic opportunities

CARBON-BASED ENERGY: THE MAIN SOURCE OF ELECTRICITY IN AFRICA

In Africa, electricity production is largely carbon-driven. In 2017, the proportion of fossil fuels (oil, gas, coal) in the different electricity production sources (the electricity mix) was more than 81%. However, this composition very much reflects the mixes found in the countries of North Africa and South Africa. These two countries account for the installation of 76% of the electric capacity of the continent and their fossil fuel electric energy production represented 87% of the continent's output in 2017. Elsewhere in Africa, the share of fossil fuels is lower, representing 45% of the electricity mix on average, with the other half coming mainly from hydroelectricity. The share of renewable energies (solar, wind, geothermal, etc.) remains lower than in the rest of the world (about 2.5% of the mix, compared to 8% in the rest of the world).

However, the African countries are seeking to take this road by exploiting their natural potential. In recent years, North Africa and South Africa have invested massively in solar and wind energy, while East Africa is beginning to exploit its geothermal potential, with Kenya leading the way. With these different strategies, electricity mixes are becoming increasingly diversified in the different regions (see map).

SOLAR ENERGY AS AN OPPORTUNITY FOR AFRICA'S ENERGY TRANSITION

With almost 600 million people living without electricity and strong demographic growth, Africa must continue to increase its energy production capacities (see pp. 16–17). Among the available options, solar energy represents genuine potential on account of the continent's comparative advantages: the yields of solar installations here are more than twice as high on average as in Germany. The massive reduction in the cost of photovoltaic solar energy facilities, for which the installation cost was almost halved between 2012 and 2016 in Africa, can represent an economic opportunity for the countries on the continent: although more expensive in terms of capital costs than fossil energy power stations, solar plants are more profitable over the long term, as their "fuel" is unlimited and comes free of charge.

SMART GRIDS, TOOLS FOR THE ENERGY TRANSITION IN AFRICA

The introduction of solar energy in Africa does imply one considerable challenge for electricity producers on the continent: unlike the other sources of energy, the flow of solar energy is intermittent by nature, as it varies according to the times of day and weather conditions. This complicates the management of electricity grids that operate on a just-in-time basis: solar electricity must therefore be used as a complement to other energy sources in order to guarantee continuous supply. African electricity distribution grids must also be upgraded to manage this technical issue.

So-called smart grids do offer a solution: these systems aim to automate the management of energy sources according to the availability of each of them. Optimized steering of the supply limits the risks of power cuts, voltage drops or surges, and helps to locate areas where there are problems in the supply (accidents, fraud, etc.). These improved networks can also offer complementary services that enable grid management to be enhanced. The introduction of smart meters for consumers, which are capable of producing a detailed analysis of their use, allows better monitoring of demand and needs, meaning that production volumes can then be adapted.

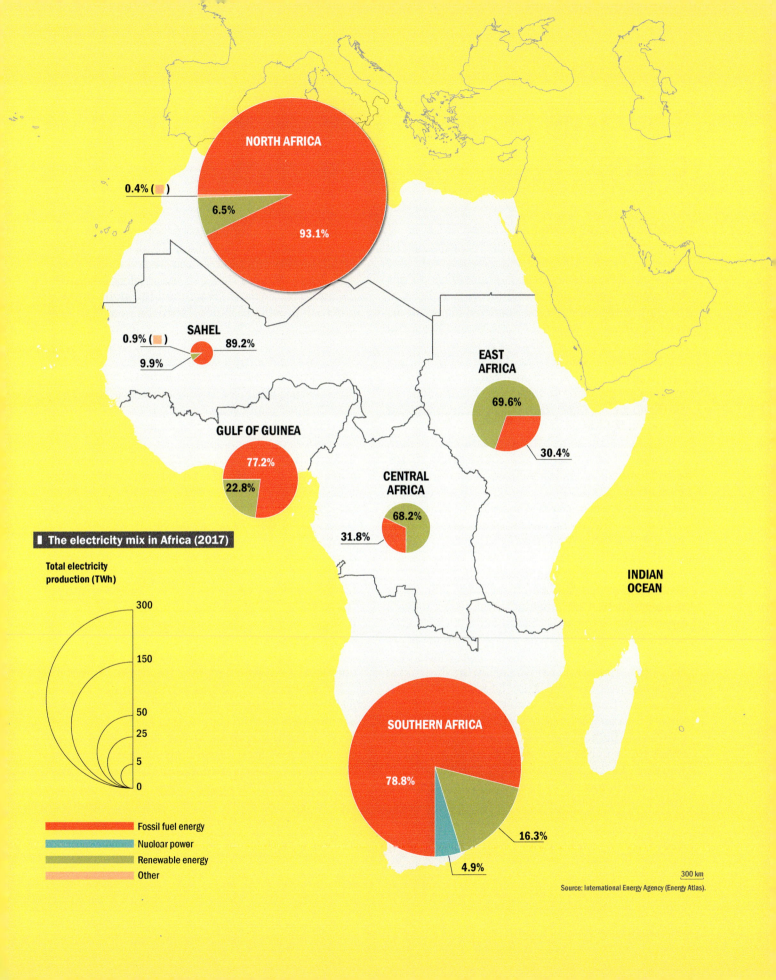

Towards more inclusive African cities

INFORMAL URBANIZATION DOMINATES

On account of the fast pace of urbanization on the continent and the still-limited income levels of many countries, African cities are expanding in conditions that are often far from satisfactory. In 2014, 46% of the continent's urban population lived in precarious neighbourhoods, compared to 30% around the world (see map). Africa is home to some of the world's most populated slums, such as that of Kibera in Kenya, just 5 km from the centre of Nairobi (population of at least 700,000 people) and the township of Khayelitsha in the urban area of Cape Town in South Africa (at least 400,000 inhabitants).

Even though the share of the population in question is declining as a proportion (it was 54% in 1995), the phenomenon is not being reduced. With the acceleration in urbanization—1.53 billion urban-dwellers by 2050—these precarious neighbourhoods will remain an important form of housing in Africa, and the number of people living there is going to grow.

THE URBAN PLANNING CHALLENGE

Africa's precarious neighbourhoods were built spontaneously on unused land and therefore are not well integrated into the urban space and are rarely connected to the core networks and grids (transport to the workplace, electricity grids, piping networks, etc.). They thus concentrate a large part of the difficulties linked to poverty and the lack of basic amenities: limited access to electricity, water and sanitation. In Angola, for example, only 20% of the one million people who live in the district of Cucuaco in Luanda have access to the public water supply, and only 30% to a system of weekly waste collection. This situation is caused by the absence of urban planning and also by chronic under-investment in urban infrastructures and in housing. Some countries have introduced policies to transform these neighbourhoods: through social housing programmes and the granting of property deeds to those living in informal housing, the countries of North Africa reduced the share of their population living in slums from 31% in 1995 to 12% in 2019.

FROM THE DEMOLITION OF PRECARIOUS NEIGHBOURHOODS TO THEIR REHABILITATION

Faced with urban precarity, one of the responses that prevailed for many years in Africa was the demolition of informal housing. This approach proved to be ineffective—the displaced population just settled in other precarious neighbourhoods—and has gradually been dropped in favour of attempts to rehabilitate the districts in question. For example, within the framework of the Participatory Slum Upgrading Program (PSUP) launched by UN-HABITAT in 2008, 25 African countries have put an end to forced displacement as a means of addressing precarious neighbourhoods. The new approach is focused on consultation between civil society in the districts in question and the public authorities, and on the development of urban amenities managed by the communities.

In Madagascar, for example, in the most precarious districts of Antananarivo, the capital, different facilities (tarmac roads, water fountains, sanitary facilities, washhouses and waste containers) have been installed by the state as part of the "Lalankely" project. Their management has been entrusted to committees of local inhabitants, in order to make sure that the projects are based on the existing social ties in these districts and that they correspond to the population's needs over the long term.

People walking on the steps of Antananarivo, Madagascar. Photo © Pierre-Yves Babelon/Shutterstock

Urban population living in slums (2014)

Share of the population living in slums (as a % of the urban population)
- 75%
- 50%
- 25%
- ND

Source: Reporters Without Borders.

Expanding internet access to accelerate the digital revolution

THE INTERNET: AFRICA'S NEW FRONTIER

Although mobile telephone systems are well established in Africa, the expansion of the internet is progressing less rapidly: while 81% of the population have a mobile telephone (see pp. 18–19), only 30% were connected to internet in 2017. The share of internet users is half that observed globally (50%) and conceals significant disparities on the continent: half of the inhabitants in the countries of North or South Africa used the internet in 2017, compare to less than one in 20 in Niger or Burundi. The material possibility of internet access is the main obstacle, relying on the existence of "network" infrastructures or access to a digital device (a connected computer or telephone).

The high price of broadband connections is another limiting factor: in 2018, one gigabyte of data cost $9 on average on the continent, which is three times higher than in France, for example. The need to master one of the main languages of the internet (English, French, etc.) and/or low levels of literacy can also be additional obstacles.

THE DIGITAL SOVEREIGNTY CHALLENGE IN AFRICA

In Africa, the challenge of internet access is compounded by the need to reinforce the digital sovereignty of the countries on the continent, as access to the internet is no doubt insufficiently secure there.

The first difficulty is that the physical internet infrastructures (data centres, servers, etc.) are rarely based on the continent. African governments and companies are therefore often forced to host their services and data outside their national territory. The continent is also lacking in back-up solutions in the event of a breakdown of the main equipment, piracy or cyber-attacks: most of the countries which face severe to moderate disconnection risks are located in Africa. Finally, personal data currently enjoys a low level of protection: only 14 countries have written texts guaranteeing personal data protection into their law.

PROPOSALS FOR ROLLING OUT THE INTERNET AND ITS USES IN THE TERRITORIES

In order for the internet to realize its potential in matters of inclusion and development for the continent, Africa will have to take up two challenges.

The first is that of deploying broadband infrastructures at the three levels of the grid: at the national level (undersea or land cables connecting the country to the worldwide network); at the domestic level (connections to the secondary cities of the country and neighbouring countries); and at the local level (distribution to users). There are off-grid innovations facilitating the connection of remote rural areas, such as internet via satellite. However, these may only be one-off solutions (military bases, bush hospitals, etc.): their cost is likely to remain higher than that of internet via optic fibre for a long time yet, and therefore cannot serve the greatest number of people. The development of a wired network therefore seems essential in the long term.

The second challenge is that of generalizing the ability to use the internet, which means developing digital learning among the population as a whole. The roll-out of the internet can therefore worsen inequalities between those who are capable of mastering this new tool and others who are often less advantaged—people living in rural areas or precarious neighbourhoods—who are less able to benefit from it owing to the lack of digital skills.

Internet users on the African continent

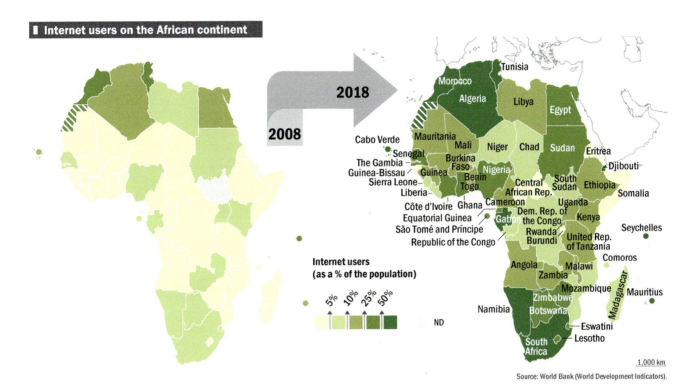

Internet users (as a % of the population): 5%, 10%, 25%, 50%, ND

Source: World Bank (World Development Indicators).

Regional differences in internet access

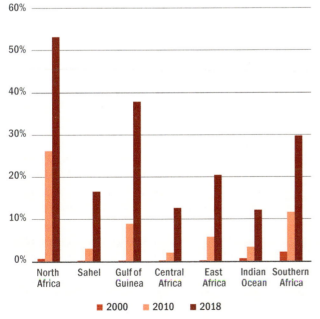

■ 2000   ■ 2010   ■ 2018

Source: World Bank (World Development Indicators).

The "Redd + La Mé" project in Adzopé in Côte d'Ivoire is attempting to halt the processes driving land clearing, while making a significant improvement to living conditions for farmers and the population as a whole. It is based on GeoPoppy, a geographical information system hosted on a portable mini-server and which can be displayed on a tablet owing to the touch screen. It can be used to track precisely any changes in cultivated and forest land areas.

Photo © Sia Kambou/AFD.

Building the capacity for an active public sector

AFRICA IN A NEW PHASE OF DEBT ACCELERATED BY COVID-19

After a first cycle of debt, which began in 1970 and was brought to an end in the 2000s by significant debt write-off programmes, Africa has embarked upon a new phase of debt. The average level of government debt—which had been brought down from 80% of GDP in 1994 to 29% in 2008—was back up to 58% of the continent's GDP in 2019. This level is close to the average for the emerging and developing countries, but the rise in this public debt makes loans more costly and less easy to come by, as the repayment capacity of the African states is perceived to be lower than that of other countries. In addition, the rise in government debt will no doubt be accelerated further by the Covid-19 crisis.

The structure of the debt has also become more complex, as the creditors have become increasingly diversified: in addition to the Western or international donors (the World Bank, the IMF), a large share of the debt is now owed to emerging countries—China in particular—but above all to international or local private actors. The debt of private actors on the continent (households, companies, etc.), meanwhile, remains well below levels observed elsewhere in the world.

DOMESTIC RESOURCES OF THE STATE AS A MEASURE OF ITS CAPACITY FOR ACTION

The capacity of states to raise taxes and duties is also of particular importance, as it largely determines their possibilities in terms of taking on debt and investing. Many African countries, however, are in a fragile situation: in half of them, government revenue as a proportion of GDP is below 20%. This situation is caused primarily by structural factors, such as the informal nature of the economy or weak administrations. On average, the level of revenues is lower in the poorest countries, and higher in the richest countries on the continent (North or Southern Africa).

In certain countries, state revenues are largely reliant on the continent's oil or mining resources, which therefore exposes them to sharp ups and downs (see p. 20). This situation is not set in stone, however, and some states—Eswatini, Ghana, Rwanda, Kenya and Uganda—have succeeded in increasing their level of revenues by several percentage points of GDP since 2010.

REINFORCING TRANSPARENCY AND ACCOUNTABILITY IN PUBLIC FINANCES

These various issues echo the efficiency of the state and the confidence users have in the administration. To enhance these two aspects, many African states have made the way their budgets are prepared and implemented more transparent. Over the past decade, they have carried out profound reforms in what is referred to as "management by performance". The aim is to improve accountability as to the efficiency and quality of public services, with the states reporting on their budget choices against public policy objectives that are defined very precisely in advance. The expected benefits of this type of reform are many, such as greater consent to taxation among people, more effective monitoring of public spending, or enhanced debt management, thereby easing access to financial markets.

These reforms have already produced results, and improvements have been made over the past two decades, as shown by the last Open Budget Survey (2017). This progress remains fragile, however, and needs to be consolidated in different ways: more systematic publication of the information required for public participation, and the reinforcement of the role of parliaments and supervisory bodies, such as the Courts of Auditors, are two possible avenues.

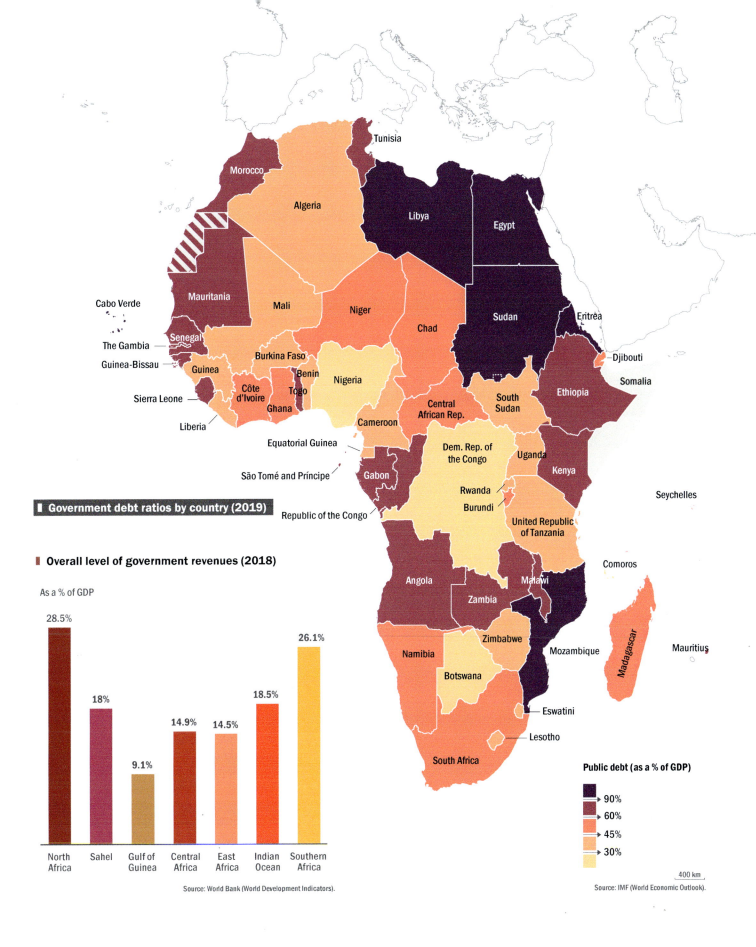

The potential of the public development banks in Africa

LITTLE-KNOWN ACTORS IN FINANCING DEVELOPMENT

The public development banks are a little-known actor in the international financial architecture, and yet they finance about 10% of world investment each year. These banks of various sizes and scopes of action—local, national, regional or global—stand out from the other financial institutions owing to the fact that their capital is controlled by a state and that they have a purpose linked to the public interest. Their choice of financing is not guided solely by profitability and they also promote strategic or sectoral objectives in order to implement the public policy of their state. They invest, sometimes on more preferential terms than those of the market, in projects that are of little interest to the commercial banks on account of their risks or the low profitability of the activities in question: small companies, rural infrastructures, social housing, R&D projects, etc. There are almost 45 public development banks in the world.

Africa is home to almost one-fourth of them, but their operational capacities remain limited compared to those in other regions: in 2018, the 103 establishments identified on the continent held 1.3% of the global assets of this family of institutions, compared to 44% for Asia and 6% for Latin America.

PUBLIC DEVELOPMENT BANKS WIDESPREAD ON THE CONTINENT

In Africa, there are public development banks in almost all the states. Two multilateral banks operate at a continental level: the African Development Bank (AfDB), headquartered in Côte d'Ivoire, and Afreximbank, based in Egypt. The significant actors also include sub-regional public banks, whose remit is to foster regional integration. This is the case, for example, of the West African Development Bank and the Development Bank of the Central African States.

At the national level, the largest public bank on the continent is the Caisse de Dépôt et de Gestion in Morocco, with assets amounting to $26 billion in 2018. South Africa, meanwhile, has seven development banks on which it can rely to implement its public policies. The majority of the other African public banks are national in their dimension and modest in size, with a volume of assets that is usually less than $500 million.

A STRATEGIC FINANCING TOOL

The public development banks could play a decisive role in the transformation of Africa. They serve as a relay to channel international financing towards African markets (see pp. 110–111) and may serve as a tool for implementing development policies. Among the eligibility criteria for their financing, they could go further in integrating indicators other than purely financial ones, relating to the Sustainable Development Goals (SDGs): protecting biodiversity, reducing territorial imbalances, including women and young people, etc.

Certain banks have already taken this path. The East African Trade and Development Bank combines objectives of regional integration with objectives of adaptation to climate change, in line with the Paris Agreement. The Development Bank of Southern Africa is supporting some 60 or so projects in raising climate financing. The Caisse de Dépôt et de Gestion in Morocco is stepping up its action in favour of the energy and ecological transition and of social and financial inclusion. Among others, it is financing new towns of the "eco-city" type. On the whole, the African public development banks must make a shift towards more rigorous governance and endow themselves with decision-making systems to assess the alignment of their financing activities with the SDGs.

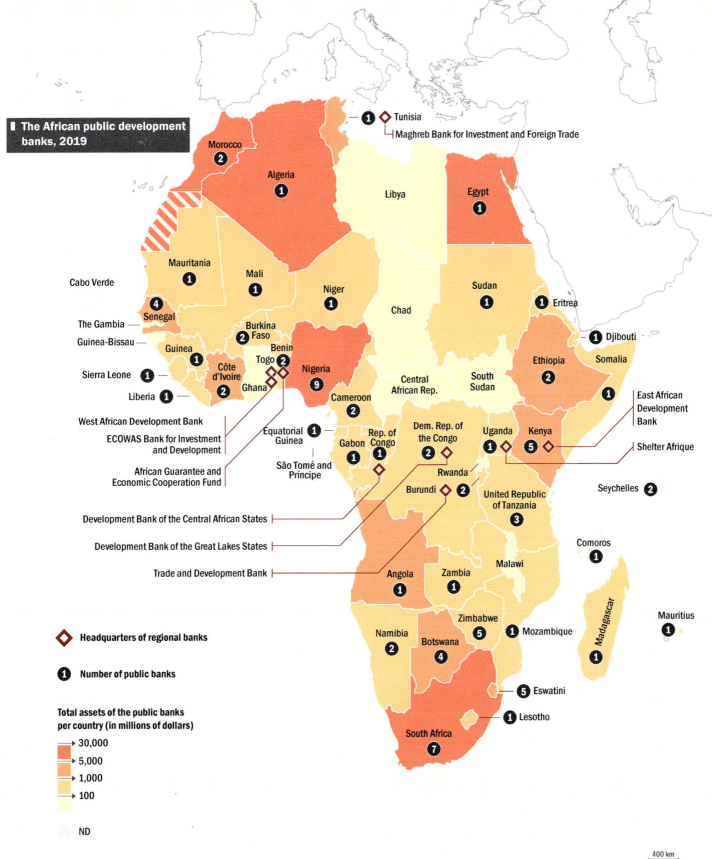

Towards more bank financing for the private sector

CONSOLIDATING THE AFRICAN BANKING SYSTEM

After two decades (1980–1990) marked by severe systemic crises in many countries, the African banking sector has been profoundly transformed and cleaned up since the 2000s. The reforms carried out by states, such as the introduction of stricter prudential rules, have been conducive to the consolidation of local banks involved in financing the private sector. The amount of borrowing by companies on the continent thus increased from 26% of GDP in 1990 to 48% in 2018.

However, this trend reflects the size of the financial sector in a small number of countries (South Africa and certain countries of North Africa and the Indian Ocean). In a great majority of countries, it remains lower than the average rate of 41% prevailing in the world's least advanced countries.

SMES SHORT OF FINANCING

The African banks prefer to lend to the lowest-risk actors: the public sector for large infrastructure projects or cash flow requirements, and the large local or multinational companies whose risk profile makes them eligible for credit. However, too few micro, small and medium-sized enterprises (MSMEs), of which there are almost 50 million on the continent, representing over 90% of the entrepreneurial fabric and 60% of employment, have access to bank financing. And yet their needs are considerable: they were estimated to be $510 billion in 2018, while the actual amount lent to them that year was $94 billion, equating to a financing shortfall of $416 billion.

This shortfall can be explained notably by the poor match between the credit solutions offered by the commercial banks and demand which is often informal, scattered and perceived as high-risk: why and how should loans be granted to individuals or companies when it is difficult to estimate their repayment capacity, the authenticity of their revenue and taxation figures, or their ability to implement a growth plan? This is a major obstacle to development in Africa, where MSMEs are necessarily going to have to play a key role in generating wealth.

FINANCING SOLUTIONS TAILORED TO INFORMAL CONTEXTS

Aware as they are of the potential of African MSMEs, the financial institutions are in search of solutions to get around these risks or mitigate them. One solution is to engage the resources of public development financing institutions which are less reluctant to bet on projects targeting objectives other than purely financial ones, in the SME financing circuit, in the form of credit guarantees granted to private banks, which are thus relieved of some of the risks.

And other mechanisms are also being invented to reduce the risks and reach a greater number of MSMEs: financing accompanied by monitoring and guidance to reinforce the entrepreneurial capacities

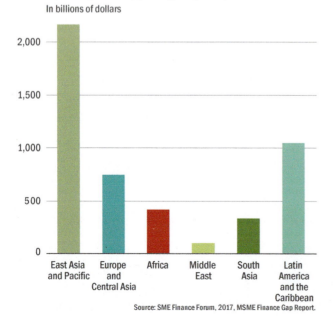

Shortfall in financing for micro, small and medium enterprises (MSMEs) per region (2017)
In billions of dollars

Source: SME Finance Forum, 2017, MSME Finance Gap Report.

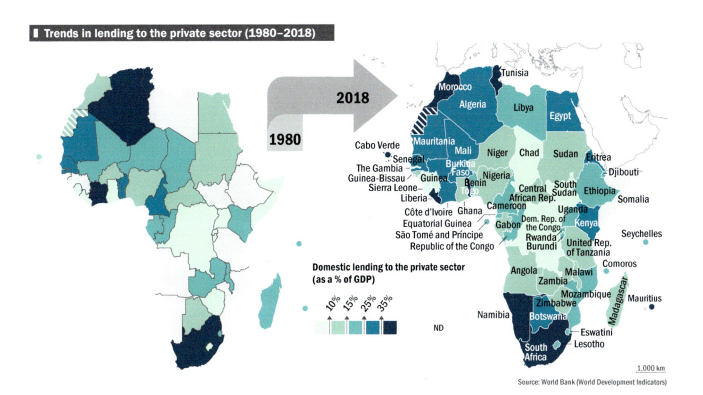

Trends in lending to the private sector (1980–2018)

Domestic lending to the private sector (as a % of GDP): 10%, 15%, 25%, 35%, ND

Source: World Bank (World Development Indicators)

of the beneficiaries; leasing, involving the granting of a loan for the purchase of an asset (a truck, for example), which then serves as collateral if the borrower should fail to repay the debt (the truck can be seized by the bank); or digital solutions developed notably with fintechs on the basis of mobile phone payments (see pp. 18–19), with a view to collecting data on the behaviour and financial capacities of the borrowers.

The effectiveness of such solutions does depend, however, on the institutional context and the profile of the beneficiaries, and it is most certainly a combination of various "bespoke" mechanisms that will be required to boost financing of the private sector in Africa.

This young woman created the Art Deco company designing and building furniture in 2014 in Médenine in Tunisia. She received a bank micro-loan via the Initiative Médenine and the Tunisian Solidarity Bank (BTS) to launch her business. The programme has facilitated access to financial resources for first-time business owners and young entrepreneurs in the Médenine region.
Photo © Augustin Le Gall/AFD.

Reinventing external financial support for Africa

REMITTANCES FROM MIGRANTS: A MAJOR SOURCE OF EXTERNAL FINANCING FOR AFRICA

In addition to foreign direct investments (FDI) (see pp. 26–27), Africa also receives two main types of external financing: migrant remittances and foreign aid.

The remittances from migrants to their country of origin continue to grow and dominate the other components of capital flows. They represented $83 billion in 2018, which is over 40% of the external financing of the continent. The sums come mainly from Europe and Asia (see map) and the main beneficiaries of these remittances are Nigeria (32%), Egypt (29%) and Morocco (11%) (see graph). The remittances from migrants are therefore essential to finance development and reduce poverty, especially as the money goes directly to the families.

FOREIGN AID STILL OF KEY IMPORTANCE

The volume of international public contributions to financing Africa fell sharply in the 1990s, before rising again from 2000 onwards, to reach $60 billion in 2018 (see graph). With their strong economic growth, some countries have automatically seen a reduction in the weight of these foreign contributions, such as Ethiopia, Uganda and Rwanda, where the volumes received since 2000 have fallen in a marked manner as a proportion of their GDP.

However, this external resource remains of capital importance for many countries, especially those in a fragile situation and which count more than the others on official development assistance, which represented as much as 33% of GDP in Somalia, and over 34% of GDP in South Sudan in 2018.

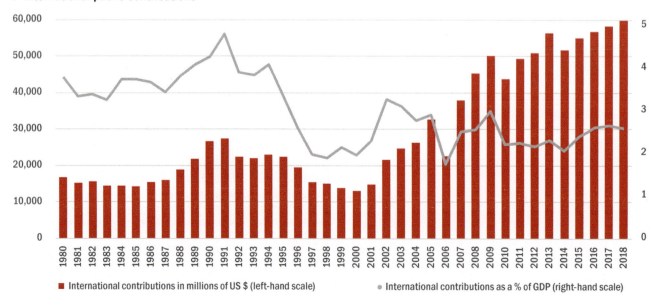

Source: World Bank (Migration and Remittances Data).

PROMOTING SUSTAINABLE DEVELOPMENT INVESTMENTS (SDI)

In Africa as elsewhere, the Sustainable Development Goals (SDGs) will only be achieved if the investments that are made are compatible with them: this concept is what is referred to as sustainable development investment (SDI). However, in order to promote SDI among private or institutional investors, it does actually need to be defined, and that is what the international community is working on by defining investment rules, principles and standards that are either general (whatever the activity) or sector-specific (for specific activities).

For instance, the UN "Principles for Responsible Investment" set out guidelines for investors to ensure compatibility of operations with the general objectives of world society in environmental, social and corporate governance matters. There are now equivalent principles for the banking community (Principles for Responsible Banking) and for insurers (Principles for Sustainable Insurance).

The Extractive Industry Transparency Initiative (EITI) is more targeted and defines standards for the exploitation of natural resources, notably by promoting the publication of annual reports and disclosure of the companies in the sector. These initiatives overlap—there are several dozen of them covering a variety of sectors, activities and operational stages—but they are all contributing to gradually defining a consensus around a trajectory with the SDGs as their common goal.

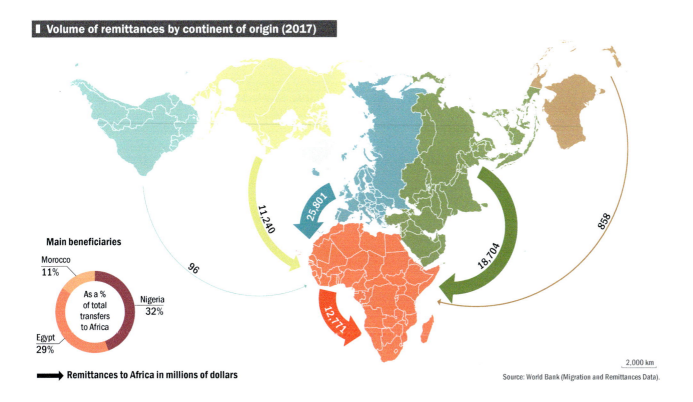

Agriculture the African way

HAS AFRICA REALLY FAILED IN ITS AGRICULTURAL TRANSITION?

Africa has experienced agricultural growth that has enabled it to feed its population better: between 1990 and 2016, per capita food production increased by 14.1%. This progress has been achieved more through the extension of the cultivated land area farmed mainly by families, made up of small farm operations (80% cover less than 2 hectares), than via intensification of production as has been observed in Latin America, or in certain countries of Asia, where yields per hectare have progressed rapidly. This gap is generally explained by a technological delay in African farming which has seen little mechanization and uses little in the way of improved seeds, pesticides and fertilizers.

However, it is important to take into consideration the key characteristics of the agricultural sector on the continent.

The state of the agricultural sector in Africa is still closely linked to the many jobs it continues to provide, jobs that have been absorbed by industry and the services sector in other parts of the world. This structural transformation is currently underway in Africa. The sector thus continues to occupy a fundamental role in the economy and social balance of the rural world. Agricultural development in Africa would therefore appear to be in line with the dynamics of transformation and settlement of the continent (see pp. 54–55).

AVOIDING THE PITFALLS OF CONVENTIONAL INTENSIFICATION

For Africa, increasing agricultural production remains an issue for food security: the continent imported 14% of its food needs in 2015 and might need to import 25% of them by 2025. One of the ways being suggested to achieve this is the extension of farming, notably by taking over vacant land to develop intensive crop farming. However, this proposal could come into conflict with the property practices of local communities, whose customary use of the land is often structured without regard for administrative ownership—in other words, in Africa, land that is officially vacant is not necessarily available, and carries the risk of destabilizing the traditional structures or increasing deforestation. Another solution that is being considered is the expansion of practices such as mechanization, the use of chemical inputs or irrigation, as the potential of the continent is underexploited in certain regions.

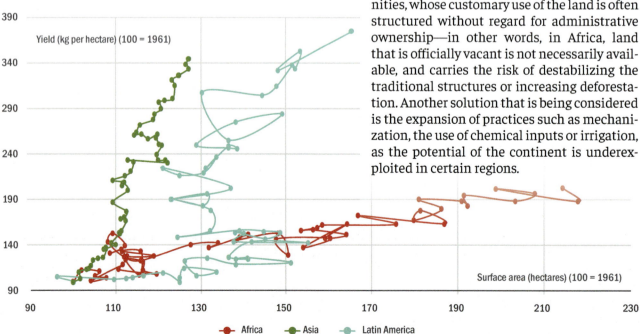

Cereal yields and land area dedicated to cereals (1961–2017)

Source: Food and Agriculture Organization of the United Nations (Aquastat).

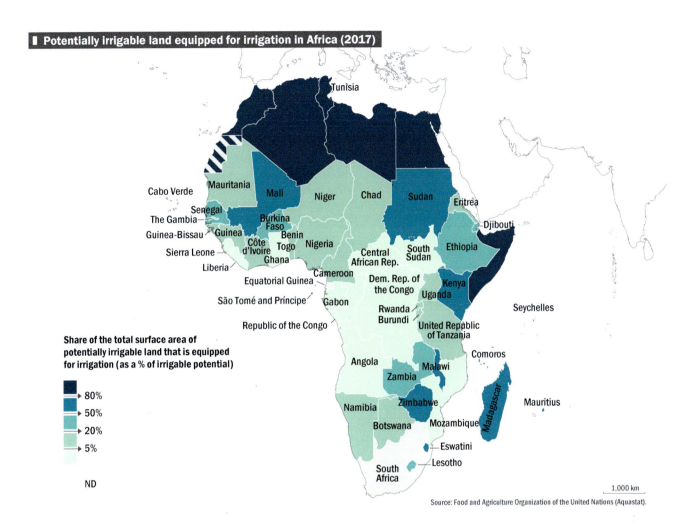

Potentially irrigable land equipped for irrigation in Africa (2017)

Share of the total surface area of potentially irrigable land that is equipped for irrigation (as a % of irrigable potential)

- 80%
- 50%
- 20%
- 5%

ND

Source: Food and Agriculture Organization of the United Nations (Aquastat).

These intensive techniques must be promoted with precaution, however: their introduction can worsen the ecological threats (water stress, pollution) and bring an early end to one of the social functions of agricultural activity: providing employment for the rural labour force.

WILL THERE BE A FARMING REVOLUTION IN AFRICA?
The agricultural transition of Africa could be achieved without suddenly forcing the farming sector off its particular path, which is to say without reproducing the intensive agricultural models developed in other contexts. The structural specifics of the African agricultural sector—maintaining a large rural population and fragile natural resources—require progressive transformations: small-scale mechanization; decentralized irrigation where necessary; the development of agro-ecological production systems (livestock rearing associations, cropping associations, etc.); and the structuring of quality systems. These are all possible options to be explored in pursuing the modernization of the sector. The variety of rural situations means that a one-size-fits-all solution will not be possible: peasant farming will remain a dominant model in the rural world of Africa, and it is very much on a case-by-case basis, working on the issues in the locality and alongside the younger generations, that the continent's green revolutions will have to be launched.

Industrialization: drawing benefit from value chains

A CURRENT DISADVANTAGE IN GLOBAL VALUE CHAINS

Most of the manufactured goods produced around the world are not made by a single company, but by a network of actors involved in different stages in the process: raw materials, logistics, transformation, marketing, etc. These fragmented production systems form "value chains" that become global when they involve actors on at least two continents.

The breakdown of that value is generally to the advantage of the companies at the downstream end of the chain in the developed countries, as they have control over access to the markets and their activities are more complex, sophisticated and profitable. At the upstream end of the chain, the activities that create less value (extraction of commodities, labour-intensive manual tasks) are generally located in the developing countries which are rich in natural resources or have low-cost labour. Africa is thus integrated into global value chains as an exporter of natural resources, but much less so as a transformer of resources than other countries (see map), which goes some way towards explaining the weakness of its manufacturing industry (see pp. 24–25).

ENGAGING IN THE VALUE CHAINS

Not all the countries of Africa are in the same position as regards the global value chains: for example, North Africa is developing high-value-added activities in the technology sectors (automobiles, aeronautics, etc.), linked with the nearby industrial centres of Europe.

However, many countries take part merely as producers of primary resources (mining, agriculture, forestry), as illustrated by the example of cocoa in West Africa. The region produces almost three-quarters of the world's cocoa beans, but draws little profit from them. In the chocolate value chain, the actors at the downstream end of the chain (chocolate manufacturers) capture most of the value, while the small-scale cocoa bean producers in Africa receive less than 6% of the sale price of a bar of chocolate sold in Europe.

Developing transformation activities in Africa would allow more value to be captured locally. This strategy of shifting further up the value chain is already underway in West Africa, which handles the grinding of one-fifth of the world's total cocoa bean production. Other countries have engaged in similar strategies, such as Uganda, a producer of raw animal hides that is seeking to develop its tanning activities and the manufacture of leather goods.

Production line workers removing freshly baked biscuits from the conveyor belt in Zambia.
Photo © GCShutter/Getty images.

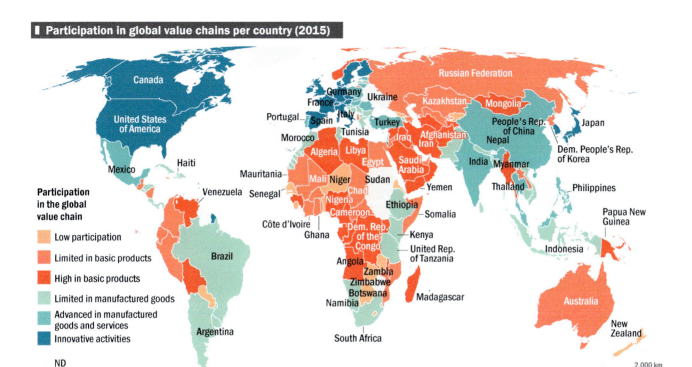

Participation in global value chains per country (2015)

Participation in the global value chain:
- Low participation
- Limited in basic products
- High in basic products
- Limited in manufactured goods
- Advanced in manufactured goods and services
- Innovative activities
- ND

Source: World Bank (World Development Report 2020).

Low participation: the share of primary goods in the total domestic added value of exports is less than 20%.

Limited in basic products: the share of primary products in total domestic added value of exports is between 20% and 40%.

High in basic products: the share of primary products in the total domestic added value of exports is equal to or greater than 40%.

Limited in manufactured goods: the rest of the sample.

Advanced in manufactured goods and services: the share of manufacturing industry and corporate services in total domestic added value of exports is equal to or greater than 80%, and that of upstream industry is equal to or greater than 30%, 20% and 15% for small, medium and large countries, respectively.

Innovative activities: income from intellectual property as a percentage of GDP is equal to or greater than 0.15% and R&D intensity is equal to or greater than 1.5% for small countries and 0.1% and 1%, respectively, for large countries.

A STRATEGIC FRAMEWORK FOR INDUSTRIALIZATION

Value chains offer a number of opportunities for industrialization in Africa. In addition to strategies consisting in shifting towards the transformation stages, another way of boosting local value is to shift upmarket, by improving the quality of local products, so as to set them apart and gain access to more profitable markets, such as the organic market or that in products of designated origin. Certification and traceability systems (fair trade, geographical indications, etc.) can be useful in this respect, as they provide a methodology for raising quality standards while providing a high profile among buyers.

Finally, another option is the rise of Africa's domestic markets (see pp. 22–23) and growing possibilities of substitution for imports, given that imports of consumer goods from the rest of the world increased from $11.2 billion to $19 billion between 2009 and 2016. These emerging markets provide African industry with local, regional and even continental opportunities that could be seized through the development of value chains within the continent.

What role could digital innovation play in Africa?

INNOVATION AT THE SERVICE OF DEVELOPMENT

In Africa, digital innovation is feeding hopes of a leapfrog or surge in development. Solutions (some based on new technologies) are being developed in all sectors—agriculture, education, health, administration—and are already providing concrete responses to development challenges, such as delivering medical equipment by drone in rural areas of Rwanda, mapping urban mobility in Cairo with a view to enhancing transport networks, providing agricultural consultancy services by mobile phone in Mali, etc.

While certain innovations may have been "imported" to the continent, if all the ambitions associated with new technologies are to be fulfilled, it will require the development of solutions devised in Africa. These innovations conceived and designed in Africa can provide a more targeted response that is better suited to the difficulties identified by African entrepreneurs. They can also be spread beyond Africa and be shared worldwide.

MULTIPLE PROFILES OF INNOVATIVE ENTREPRENEURS

Against a backdrop of limited public spending on research and development (0.7% compared to an average of 2.3% worldwide), innovation in Africa is being driven by a variety of actors. The best-known digital innovation in Africa is the M-Pesa, a mobile phone-based money transfer service (M stands for mobile and *pesa* means money in Swahili), the fruit of a partnership between a UK and Kenya-held telecommunications company, Safaricom, and a Kenyan bank, CBA. The 7,000 innovative start-ups counted on the continent are also attracting growing interest: in 2019, the most mature of them raised more than €2 billion from investors all over the world, up by 74% on the previous year. Some of the more structured start-ups are already successful: Jumia Group, the leader in African e-commerce based in Nigeria, is the continent's first "unicorn", valued at over $1 billion on the New York Stock Exchange. Most African digital entrepreneurs are very much the artisans of innovation, sometimes in precarious positions. Despite the potential of some of their initiatives, they are facing the same difficulties as other entrepreneurs on the continent (see pp. 86–87) and their activities are not always recognized socially.

INNOVATION PLATFORMS PART OF A LOCAL ECOSYSTEM

An ecosystem is also developing to support Africa's innovators and increase the likelihood of success. The number of tech hubs, incubators and business accelerators is growing quickly (up by 97% between 2016 and 2018, from 314 to 618). These support platforms are appearing all over Africa, but are mainly concentrated in a few centres of innovation: in 2019, five cities (Cape Town, Lagos, Johannesburg, Nairobi and Cairo) were home to almost half of the continent's start-ups (see map). These entities incubate promising start-ups and help them to grow (development plan, business models, etc.) before putting them into contact with their network of financial and industrial partners.

The African tech hubs often establish partnerships with the United States, external partners or private operators—especially mobile operators—to give them financial sustainability. However, if they are to reconcile innovation and development to the full, these structures face a twofold challenge: detecting and supporting innovators outside the continent's main urban centres where they are based, and adapting to the specific constraints that face African start-ups, which sometimes take longer to emerge. Sharing know-how and promoting solutions focused on mutual assistance and inclusion are two avenues to be explored in greater depth.

Index

PAGE NUMBERS IN BOLD REFERENCE TABLES.

2050 Africa's Integrated Maritime Strategy (AIM Strategy) 70

ACTIF project 21
Afrex-imbank 106
Africa-European Union summits 30
African common passport 66
African Continental Free Trade Area (AfCFTA) 23, 59
African Development Bank (AfDB) 22, 62, 66, 106
African Union 30, 58, 59, 70
African Union Commission (AUC) 66
"African Water Vision for 2025" 70
agricultural sector 24, 112–113
air pollution 78
air transportation industry 28, **29**
Al Bawsala organization 94–95
Algeria 32, **41**, 50
Amazonian forest **74**
Angola **41**
Arab Maghreb Union (AMU) 58, **59**
armed conflicts 34–35, **35**, 49
Asia 8

Banc d'Arguin Reserve 82–83
Bangladesh **21**
banking services: mobile telephones and 18, **19**, 109; for private sector 108–109; public development banks 106, **107**; see also financial services
banking systems 62, **63**, 108–109 see also financial industry
biodiversity 82–83
Botswana 44, 50
boys see youth
Brazil 30
"Bruits de Tambours" project (Senegal) 92
Burkina Faso 49
Burundi 16, 97, **97**
business reforms 36, **37**

Caisse de Dépôt et de Gestion (Morocco) 106
Cambodia **21**
cartoon strips 69
CBA 116
cellular phones see mobile telephones
Central Africa **3**, **4**, 47
Central African Economic and Monetary Community (CAEMC) 58
Central African Republic 35, **35**, 49
cereal yields **112**
Chad, Maison de la Petite Entreprise **87**
Chad Health Sector Support Project (PASST2) 5
child mortality 8, **9**
children see youth
China, People's Republic of **21**, 26, 30 chronic diseases 90–91
cities see urban spaces
civil society 94–95, **95**
climate change: CO_2 emissions 78, **79**; displaced persons and 78; ecosystems deterioration 82; greenhouse gases 78, **79**; impacts of 80; rising sea levels 80, **81**; temperature trends 78, **79**; urban cities and 80; water stress 80, **81**
cocoa bean production 114
commodities 44
Common Market for Eastern and Southern Africa (COMESA) 58, **59**
communication technology 18
Community of Sahel-Saharan States (CEN-SAD) 58, **59** [we have this with em-rule or possibly en-rule – please make sure pages 58 and 59 have hyphen]
Congo Basin forest 74, 75, **75**
conurbations **53**
Côte d'Ivoire **21**, **103**, 106
"countercyclical" policy xii
countryside see rural spaces
Covid-19 crisis xi–xii, 10, 20
cultural and creative industries 68–69
cultural cooperation agencies 30, **31**
customs unions 58

Dar es Salaam, United Republic of Tanzania 52
deforestation 75
Democratic Republic of Congo: armed conflicts 35, **35**; Kinshasa 52; megacities 52; refugees 49
democratization: civil society and 94–95; human rights and 94; media's role in 92; at political level 32, 92; rule of law and 32
demographic dividend 7
Development Bank of Southern Africa 106
diabetes 90, **90**
digital entrepreneurs 116
digital innovation 116
digital learning 102
digital technology 18, 102, **103**, 109
displaced persons 48, **49**, 72, 78, 100 *see also* refugees
drinking water 14–15, **15**

East Africa: fertility rates **4**; migrants/migration 64; population growth 2; population trends **3**
East African Community (EAC) 58, **59**, 66
East African Trade and Development Bank 106
East Asia 12
Economic Community of Central African States (ECCAS) 58, **59**
Economic Community of West African States (ECOWAS) 58, **59**, 66
economic cooperation, intergovernmental organizations for 58–59
economic development 24, 44, 96
economy: agriculture and 24; commodities and 44; Covid-19 crisis and 20; daily income distribution **23**; expansion of 22; farming sector 24; GDP per capita 20; growth of 20–21, **21**, **22**; industrialization and 24; internal market 22, 23; large economies 40; macro-economic improvements 36; manufacturing sector 24; performance 40; population effects on 40; public governance 36; regional economic integration 58–59; small economies 40; STEM teaching 84–85; tourism 28; women's role in 88; *see also* trade
education: access inequalities 13; ADEM project 85; completion rate 84, **85**, 97; GDP devoted to 12; government spending on 12, **13**; qualified teachers **84**; school durations **13**; school enrolment 12, 13, 84; school enrolment rates 12; sports and 97; teacher training 84–85; teaching conditions 84

Egypt: female entrepreneurs 88; GDP per capita **41**; income inequality 50; public development banks 106; tourism 28
electricity: access obstacles to 57; access to **17**, **56**, **57**; electrification levels 16; primary schools with access to **84**; production 98; production mix **99**; rural vs. urban access to **56**; solar power 17; *see also* solar energy
electricity grids 98
electrification **17**
employment: under-employment 86; entrepreneurship and 86–87; farming sector 24; quality 86; services-related 24; unemployment insurance coverage **86**; unemployment rates 86; women's workforce participate rate 88; youth and 86, **87**
endangered species 83
endemic diseases 10
Endowment Fund for the Banc d'Arguin and Coastal and Marine Biodiversity 83
energy, carbon-based 98 *see also* solar energy
entrepreneurship 86–87, 88, 108–109, 116
epidemics 10
Equatorial Guinea 42
Ethiopia **21**, **41**
Europe 28
exports *see* trade
Extractive Industry Transparency Initiative (EITI) 111
extractive sector 24
extreme poverty 46–47, **46**, **47**
Ezulwini Consensus 30

farming sector 24
female entrepreneurship 88
female genital mutilation 88, **89**
females *see* women
fertility rates 4, **4**
film production 68, **69**
financial industry: banking capitalization **63**; banking systems 62; development of 62; market capitalization **63**; private equity funds 62; regional financial centres 62; stock markets 62
financial services: digital technology for 18; for micro, small and medium enterprises (MSMEs) 108–109; migrant remittances and 110, **111**; mobile telephones and 18, **19**; for private sector 108–109, **109**; *see also* banking services

fish migrations 70
food production 112
food security 112
foreign aid 110, **111**
foreign direct investments (FDIs) 62, 110
forests 74–75
Forum on China-Africa Cooperation (FOCAC) 30 [we have this with en-rule in text on page 30 – please change to hyphen]
fragility and development 48–49, **48**
France-Africa summits 30
freedom of information 92
free movement systems 64, 66
free trade areas 58, 59

Gabon 42
GeoPoppy **103**
Ghana **41**, 69
Gini Index 50
girls *see* women, youth
Global Alliance for Vaccines and Immunization (GAVI) 8
global value chains 114, **115**
global warming 8, 78
government *see* public sector
Green Climate Fund (GCF) 78
greenhouse gases 78
Gulf of Guinea: democracy, demand for 32; fertility rates **4**; migrants/migration 64; population growth 2; population trends **3**; poverty rates 47; school enrolment 13

Hague Declaration on Equal Access to Justice for All 92
happiness and well-being: measurements of 50; perception of **51**
health and development: chronic diseases 90–91; diabetes 90, **90**; endemic diseases 10; fertility rates 4–5; financing of 91; healthcare systems, shortcomings of 10; HIV prevalence rates **11**; infectious diseases 10; life expectancy 4; malaria 10; social dimension of 10; tropical diseases, persistence of **11**; vaccinations 8; water and sanitation 14–15
health systems 91
Heavily Indebted Poor Countries (HIPC) initiative 36
HIV prevalence rates **11**

Horn of Africa 8, 64, 88
human rights 94

imports *see* trade
income inequality 50, **51**
India **21, 68**
India-Africa summit 30
Indian Ocean: fertility rates **4**; population trends **3**; poverty rates 47
industrialization: air pollution and 78; electrical connectivity and 16; strategic framework for 115; value chains and 114–115, **115**
industrial sector 24, **25**
infectious diseases 10
informal housing 100
infrastructures: access obstacles to 57; airport 28; broadband connections 102; deforestation and 75; health challenges due to 15; hydroelectric 70; industry development and 24; internet access 102; under-investment in 100; investments in 20; landline networks 18; resilient to climate change 80; rural communities and 56; Saharan-Sahelian space 72, **73**; sports 96–97; transcontinental 57; *see also* electricity
Initiative Médenine **109**
Inter-Governmental Authority for Development (IGAD) 58, **59**
Intergovernmental Panel on Climate Change (IPCC) 78, 80
International Food Policy Research Institute (IFPRI) 60–61
International Institute for Democracy and Electoral Assistance (IDEA) 94
International Monetary Fund (IMF) 61, 86
International Union for Conservation of Nature (IUCN) 82
internet access 68, 102, **103**
intra-African trade 58–59, 60–61
intra-continental foreign direct investments 62
intra-continental mobility 64, **65**
intra-continental trade 59
Ireland **21**
irrigation **113**

Jemaa el-Fnaa Square 23
Jumia Group 116

justice, access to 92, **93**

Kenya: cartoon strips 69; digital technology 18; electrification levels 16; GDP per capita **41**; *Leti Arts* 69; video games production 69
Kinshasa, Democratic Republic of the Congo 52

labour market *see* employment
Lagos, Nigeria 52
Lake Victoria 70
"Lalankely" project 100
"La Main à la Pâte" Foundation 85
Lao People's Democratic Republic **21**
large economies 40, **41**
Latin America 12
least developed countries (LDCs) 42
"Les Puits du Désert" association 95
lethal conflicts 34–35
Libraries Without Borders 7
Libya 35, **35**
life expectancy 4, **5**

macro-economic improvements 36
Madagascar 100
malaria 10, 15
Mali crisis 72
Malta **21**
management by performance 104
manufacturing sector 24, **25**, **61**
Marrakesh, Morocco, Jemaa el-Fnaa Square 23
maternal mortality 8
Mauritania 82–83
Mauritius 42, **43**
media freedom 92
megacities 52
micro, small and medium-sized enterprises (MSMEs) 108–109
middle class 22
middle-income countries 42
migrants/migration: banking services 18; digital technology 18; free movement systems and 64; geographic proximity and 64; intra-continental mobility and 64, **65**; poverty and 64; remittances from 110, **111**; *see also* refugees
Miombo Forest 74
mobile telephones 18, **19**, 68–69, 102

mobility 64, **65**, 66
Mo Ibrahim Foundation 36
monetary integration 58
Morocco 28, **41**, 106
movie production 68, **68**, **69**
M-Pesa 116
Mukwege, Denis 88
Myanmar **21**

Namibia 50
natural habitats, protection of 82–83
natural resources 44, **45**, 72, 111
neglected tropical diseases (NTDs) 10
Niger, "Les Puits du Désert" association 95
Nigeria: armed conflicts 35, **35**; film production 68, **68**, **69**; GDP per capita **41**; Lagos 52; megacities 52
"Nollywood" movie industry 68, **69**
non-state conflicts 34
North Africa: fertility rates 4, **4**; income inequality 50; migrants/migration 64; population trends **3**; poverty rates 47; school enrolment 13

obesity 90

pandemics 91
Participatory Slum Upgrading Program (PSUP) 100
pediatrics package 5
pension systems 6
People's Republic of China *see* China, People's Republic of Phillipi, Cape Town, South Africa 51
piracy 70
Playdagogy Programme 97, **97**
PLAY International NGO 97
political regimes **33**
political liberalization 32
political violence 34
population: age structure of 6; daily income distribution by **23**; demographic trends **3**; density 78; displacements 78; growth 4–5; poverty and 46; of rural spaces 54; Saharan-Sahelian space **73**; surge in 2; working 7; world density **55**; world distribution of 2; youth 6, **7**
poverty: decline in 46; extreme poverty rates **47**; population growth and 46; projections of **46**; refugees and 64
premature births 5

press freedom 92, **93**
private equity funds 62
production sector 61
protected spaces 82
public development banks 106, **107**
public sector: external financing for 110; finance transparency/accountability 104; governance 36; governance, forms of 94–95; government debt 104, **105**; government debt ratios **105**; government revenues 104, **105**; management by performance 104

"Redd + La Mé" project **103**
refugees 49, 64 *see also* migrants/migration
Regional Economic Communities (RECs) 58–59, **59**, 64, 66
regional economic integration 58–59
renewable energy 16
Reporters Without Borders (RWB) 92
research and development 116
retirement pension systems 6
river network 70, **71**
rule of law 32, 92
rural spaces: development disparities with urban spaces 56–57; population density 54, **55**; *see also* urban spaces
"rurbanization" effect 52
Russia-Africa summit 30
Rwanda **21**, 84

Safaricom 116
Sahara Desert 72
Saharan-Sahelian space 72
Sahel countries: fertility rates 4, **4**; under-nutrition in 8; population growth 2; population trends **3**; school enrolment 13; *see also* Saharan-Sahelian space
sanitation 14–15
schools *see* education
sea levels 80, **81**
Senegal: ADEM project 85; annual growth rate **21**; "Bruits de Tambours" project 92
Senegal River 70
Senegal River Basin Development Authority (OMVS) 70
service economy 24
Seychelles 42
Sierra Leone 49

slum living 100, **101**
small and medium-sized enterprises (SMEs) 108–109
small economies 40
smart grids 98
social inclusion 97
social ties in Africa 50
solar energy 17, 98
Somalia: armed conflicts 35, **35**, 49; foreign aid 110; refugees 49
South Africa: GDP per capita **41**; income inequality 50; music industry 68; tourism 28
South Asia 12, 16
Southern Africa: democracy, demand for 32; fertility rates 4, **4**; income inequality 50; migrants/migration 64; population trends **3**; poverty rates 47
Southern African Customs Union (SACU) 58
Southern African Development Community (SADC) 58, **59**, 61, 66
South Sudan: armed conflicts 35, **35**; foreign aid 110; refugees 49
sports: actors in **96**; benefits from 97; economic development and 96; social inclusion and 97
stock markets 62
sub-Saharan Africa 47, 88
Sudan 35, **35**, 49
sustainable development investments (SDI) 111

Tajikistan **21**
tech hubs 116, **117**
technology: digital entrepreneurs 116; internet access 68, 102, **103**; mobile telephones 18, **19**, 68–69, 102, 109; tech hubs 116, **117**; *see also* mobile telephones
terrorism 34–35
TICAD (Tokyo International Conference on African Development) 30
tourism 28, **29**
trade: commodities in exports **45**; cross-border 60–61; European Union **26**; exports **26**, **45**, 60–61, **61**; free trade areas 58, **59**; imports **26**, **60**; international 26–27, **27**; intra-continental 58–59, 60–61; manufacturing **61**
tropical diseases, persistence of **11**
Tunisia 28, 94–95
Tunisian Solidarity Bank (BTS) **109**
Turkish embassies, in Africa 30

Uganda 49, 61, **61**
UN Agency for Refugees (UNHCR) 49
under-nutrition 8
unemployment *see* employment
UN-HABITAT 100
United Nations (UN) 30
United Nations Development Program (UNDP) 78
United Nations Framework Convention on Climate Change (UNFCCC) 78
United Republic of Tanzania **41**, 52
United States **68**
urbanization, informal 100
urban migrants 18
urban spaces: conurbations **53**; development disparities with rural spaces 56–57; informal housing 100; megacities 52; population density **55**; "rurbanization" effect 52; second-tier cities 52; slum population **101**; world's largest cities **53**; *see also* rural spaces

vaccinations 8, **9**
value chains 114–115, **115**
video games production 69
Viet Nam **21**
violence 34–35
visa openness 66, **67**
Visa Openness Index 66

water: governance 15; groundwater resources 70, **71**; irrigable lands **113**; sanitation and 14–15
water-borne diseases 15
water stress 80, **81**
wealth per capita 20, **20**, **41**, 42, **43**
wealth production 24
West Africa 10, 64, 114
West African Economic and Monetary Union (WAEMU) 58
wetlands 82
women: economic role of 88; entrepreneurship 88; genital mutilation 88, **89**; inequalities between men and 88; in labour market **89**; pre-natal care for 8; suffering sexual violence 88; workforce participation rate 88
working population 7
World Bank 36, 78, 86
World Economic Forum 88
world population, distribution of 2, **3**

youth: child mortality 8, **9**; education access inequalities 13; education efforts 12; employment needs for 86; joining labour market **87**; population 6, **7**; unemployment rates 86; vaccinations **9**

Zagtouli photovoltaic solar power plant 17
Zambia 50

MANAGEMENT AND COORDINATION

CHRISTOPHE COTTET, Doctor in development economics, graduate of the Centre pour les Etudes et Recherches sur le Développement International (CERDI), economist in the Africa Department of the AFD

CLÉMENCE VERGNE, Doctor in development economics, graduate of the Centre pour les Etudes et Recherches sur le Développement International (CERDI), economist in the Africa Department of the AFD

EDITORIAL TEAM

MAXIME WEIGERT, Doctor in economic geography, graduate of the University of Paris 1-Panthéon-Sorbonne, partner at OXCON Frontier Markets & Fragile States Consulting

CLÉMENCE VERGNE and **CHRISTOPHE COTTET**

STATISTICAL TEAMS

YASMINE OSMAN, graduate of ENS Cachan and ESCP Europe, economist in the Africa Department of the AFD

JEANNE DE MONTALEMBERT, graduate of University Paris Dauphine

EDITORIAL COMMITTEE

RÉMY RIOUX, AFD Chief Executive Officer

RIMA LE COGUIC, Director of the AFD Africa Department

THOMAS MÉLONIO, Executive Director for Innovation, Research and Knowledge at the AFD

MAPS

GUILLAUME SCIAUX, PACHA CARTOGRAPHIE

GRAPHICS

MANOËL VERDIEL

LAYOUT

ALAIN CHEVALLIER